Making Science

MAKING SCIENCE

• • • •

Between Nature and Society

Stephen Cole

Harvard University Press
Cambridge, Massachusetts
London, England
1992

Library of Congress Cataloging-in-Publication Data

Cole, Stephen, 1941–
Making science : between nature and society / Stephen Cole.
p. cm.
Includes bibliographical references and index.
ISBN 0–674–54347–5
1. Science—Social aspects. 2. Science—Philosophy.
3. Constructivism (Philosophy) I. Title.
Q175.5.C64 1992
303.48′3—dc20
91–42192
CIP

For Sylvia Cole

Contents

Preface

In writing this book I have had two main audiences in mind: the growing group of specialists in the history, philosophy, and sociology of science whose collective enterprise is now often referred to as the social study of science, and the broader sociological community. The first group has been deeply and irrevocably influenced by the revolution which has occurred in the history and philosophy of science. This revolution, marked by the publication in 1962 of *The Structure of Scientific Revolutions* by Thomas S. Kuhn, challenged the view that science was a uniquely rational activity. Today there are still major epistemological differences among practitioners of the social study of science, but there is virtual consensus that the "traditional" view of science overemphasized rationality. This new way of looking at science has, unfortunately, had only a limited impact upon sociologists and other social scientists who do not specialize in the social study of science. Despite the appeal of relativist and subjectivist critiques to many individual scholars, as disciplines the mainstream social sciences remain committed to a positivist view of science.

Perhaps it is not surprising that among those who specialize in the social study of science sociologists have been the most radical. The dominant orientation in the sociology of science, "social constructivism," is based upon a relativist epistemological position and the argument that nature has very little, if any, influence on the development of the content of science. Rather than believing that nature determines the course of the development of scientific knowledge, adherents of this approach hold that scientists socially construct nature in the laboratory.

My own views are to the "right" of the social constructivists and to the "left" of traditional positivism, but perhaps closer to the position of the constructivists than to that of the positivists. I think of my position as being that of a realist-constructivist (as opposed to a relativist-constructivist). A realist-constructivist believes that science is socially constructed both in the laboratory and in the wider community, but that this construction is influenced or constrained to a greater or lesser extent by input from the empirical world. Instead of saying that nature has no influence on the cognitive content of science, a realist-constructivist says that nature has *some* influence and that the relative importance of this influence as compared with social processes is a variable which must be empirically studied. I do not believe that evidence from the external world *determines* the content of science, but I also reject the position that it has no influence.

An intellectual movement which takes a position at variance with existing beliefs is frequently forced into overstating its claims in order to differentiate itself from the position it argues against. Thus some Marxists have argued that the *ultimate* cause of *all* social phenomena can be traced to economic forces, and some Freudians that *all* neurosis is a result of sexual repression. Some constructivists similarly overstate their case in arguing that nature has no influence on the development of science. In the past, sociologists believed that because the content of science was ultimately determined by nature, the content of science was not a topic that they could fruitfully investigate. The constructivists therefore remain fearful that if nature is even included as one of several influences on the cognitive content of science, then social influences will be seen as secondary or unimportant. Some observers might indeed come to this conclusion, but it is now unlikely that many sociologists would.

I have two different goals for this book. First, I would like to convince those sociologists who remain positivists to pay attention to the important work done in the social study of science since 1970, which provides us with a much more complex and interesting account of how science is conducted. A positivist image of the natural sciences is based upon an out-of-date stereotype. Given that social scientists frequently try to imitate or make their disciplines more like the natural sciences, it becomes especially important to uncover any errors in our understanding of these sciences. If sociologists gain more accurate knowledge of how the natural sciences grow and change, they will be better able to understand the reasons for the persisting differences between the natural and the social sciences. For my colleagues in the social study of science I hope to illustrate some of the problems that a relativist epistemological position poses for understanding science. I also hope that the empirical work presented here will demonstrate that quantitative macro-level studies of the content of science can add to and supplement results obtained from qualitative micro-level studies.

In this book I discuss a revolution which has occurred in the history and philosophy of science and compare the "new" approach with the "traditional" approach of the positivists. I am employing an "ideal-type" image of "traditional" philosophers of science, or positivists, in order to examine the major differences between this philosophy and that which has developed since the early 1960s. There is probably no philosopher of science who has held precisely the opinions which I attribute to the "traditional" view. The positions of positivist philosophers of science are of course more sophisticated than depicted in my ideal type, and it was in fact some of these philosphers who introduced one of the main concerns of the new approach, the underdetermination of theories by data. For a more detailed discussion of the contrasting views of relativist and nonrelativist philosophies of science, the reader should see Laudan (1990).

Although I criticize in detail the work of some constructivist sociologists of science, this criticism should not be misinterpreted. For too long there has been a divide between the more traditional sociologists of science, sometimes referred to as "Mertonians" or the "North American school," and the mostly European social-constructivists. The last few years have seen substantial change in this area as the significance of the constructivist program has become more widely recognized. I believe that the best work in this school represents an important contribution to the sociology of science and deserves much greater recognition by American sociologists of science. I hope that this book will serve the cause of a *rapprochement* between relativist and nonrelativist sociologists of science.

How much I have learned from the constructivists will be particularly evident in Chapters 7 and 8, in which I re-address the question of central concern in much of my previous work in the sociology of science: Is science universalistic? My answer is now substantially different from the one put forth in my 1973 monograph with Jonathan R. Cole, *Social Stratification in Science*. In these chapters I use elements of the constructivist approach to address a paradox: How is it that statistical studies have shown relatively little evidence for particularism, when virtually all members of the scientific community are aware of many cases in which particularistic criteria of evaluation were employed? I conclude that the traditional studies failed to tap adequately particularism based on cognitive criteria and location in social networks rather than on the statuses occupied by the scientists.

Another issue on which my views and those of the constructivists are substantially the same is the question of the extent to which the level of consensus in the natural and the social sciences is similar. Both the constructivists and I, in contrast with most traditional sociologists of science, believe that there are not meaningful differences in consensus at the research frontier. We have come to this position along different roads, the constructivists out of their effort to debunk the special rationality of the natural sciences and I out of the results of a series of quantitative empirical investigations. But in considering the path I have taken the constructivists may recognize the utility of such studies for investigating some of their theories.

In this book I make a distinction between the research frontier and the core of scientific knowledge. The research frontier in any discipline consists of all new work which is in the process of being evaluated by the community. The core is the exceedingly small number of contributions which are accepted by the community as important and "true." There is a low level of consensus on frontier knowledge and a high level of consensus on core knowledge. Here, too, my approach is similar to that of the constructivists, particularly Latour (1987), who makes the distinction between science in the making and "facts." The first type of science is seen as being the work of individual scientists and contextually contingent, the second type a result of the laws of nature.

Quotations from *Science in Action: How to Follow Scientists and Engineers through Society* by Bruno Latour (Cambridge, Mass.: Harvard University Press), Copyright © 1987 by Bruno Latour, and from *Laboratory Life: The Construction of Scientific Facts* by Bruno Latour and Steve Woolgar, Copyright © 1986 by Princeton University Press, are reprinted by permission of the publishers. I am also grateful for permission to reprint or adapt Figure 1, Tables 1–4, and selected text from S. Cole, J. R. Cole, and G. Simon, "Chance and Consensus in Peer Review," *Science*, 214 (November 20, 1981), 881–886, Copyright 1981 by the AAAS; Tables 1, 3–7, 10, and selected text from S. Cole, "The

Hierarchy of the Sciences?" *American Journal of Sociology*, 89, no. 1 (July 1983), 111–139, © 1986 by the University of Chicago; Tables 9–10 from S. Cole, "Age and Scientific Performance," *American Journal of Sociology*, 84, no. 4 (July 1979), 958–977, © 1979 by the University of Chicago; and Tables 3.1, 3.2, 3.4, 3.5, 3.6, Figures 1–3, and selected text from J. R. Cole and S. Cole, "Experts' 'Consensus' and Decision-Making at the National Science Foundation," in K. Warren, ed., *Selectivity in Information Systems*, pp. 27–63, © 1985 by Praeger Publishers and reprinted by permission of Greenwood Publishing Group, Inc., Westport, Connecticut. Previous versions of Tables 5.4, 5.5, and 5.6 were presented in Stephen Cole, Jonathan R. Cole, and Lorraine Dietrich, "Measuring the Cognitive State of Scientific Disciplines," in Y. Elkana, J. Lederberg, R. K. Merton, A. Thackray, and H. Zuckerman, eds., *Toward a Metric of Science: The Advent of Science Indicators*, pp. 209–251, Copyright © 1978 by John Wiley & Sons, Inc., and are reprinted by permission of the publisher.

This book has benefited substantially from my interactions with colleagues. I would like to thank the following people for providing feedback on earlier versions of all or part of the manuscript: Jeffrey Alexander, Paul Allison, Michael Aronson, Bernard Barber, Elinor Barber, Ivan Chase, Jonathan R. Cole, Maria Misztal Cole, Elizabeth Garber, Alfred Goldhaber, Karen Knorr-Cetina, Seymour Martin Lipset, Robert K. Merton, Sharon Witherspoon, Mayer Zald, and Harriet Zuckerman. In particular, for providing detailed comments on the penultimate draft, I thank Maria Misztal Cole, Robert Merton, and Sharon Witherspoon. Elizabeth Gretz helped improve the clarity of the final version.

In thinking through the many complex issues involved, I also had the aid of two undergraduate classes and two graduate seminars at the State University of New York at Stony Brook. For the first of these seminars I was fortunate to have two colleagues sit in and participate in the class. One was my friend and colleague, the late Hanan C. Selvin; the other was Elaine Winschell, Professor of Biology at Ramapo College. Both provided helpful feedback, and Elaine served as a useful informant on how experiments in the biological sciences are conducted. In addition, I have benefited from presenting various parts of this book at seminars and professional meetings. Some of the early research leading to this book was supported by a fellowship from the John Simon Guggenheim Foundation and the Center for Advanced Study in the Behavioral Sciences. Chapters 7 and 8 were originally written while I was a visiting professor at the Center for American Studies at the University of Warsaw.

My greatest debt is to the collaborators with whom I conducted much of the empirical work utilized in this book. Most important is my brother, Jonathan R. Cole, who has been my closest colleague for many years. Originally this book was supposed to be another "Cole and Cole" production, but Jon's heavy administrative responsibilities made this impossible. I also thank Garry Meyer, Leonard Rubin, and Gary Simon for their contributions to the empirical studies we did together. Scott Feld and Thomas Phelan helped with the analysis of the data in Chapter 6. These people are, of course, not responsible for the interpretations presented here.

My last acknowledgment is to my mother, Sylvia Cole. In addition to being a mother she has been a teacher, a role model, and a friend who has been a constant source of support and encouragement.

October 1991
Setauket, New York

· · 1 · ·

Nature and the Content of Science

Science is one of the most important institutions in modern society, yet the sociology of science did not emerge as a specialty until the 1960s; even now it remains the primary research area of a relatively small number of sociologists. How can we explain the comparative neglect of this institution by sociologists as opposed to historians and philosophers? The lack of attention to science by sociologists derives in large part from the long-standing belief that the most interesting aspects of science, the new ideas and important discoveries which make up its content, are ultimately determined by nature and are not subject to social influence. Traditionally, the development of scientific ideas was seen as the province of historians and philosophers of science. Even the behavior of scientists as they went about their work of trying to discover the laws of nature was thought to be guided by a set of special scientific norms and rational procedures. The very raison d'être of such standards was to leave little room for the influence of social processes, which were viewed as interfering with the pursuit of science. Since the 1960s, however, there has been a dramatic revolution in our conception of how science works. This revolution challenges the image of science as an activity which is different from other human activities and has led to a renewed interest in the sociology of science.

This book examines the ways in which social processes and variables influence the growth of scientific knowledge. It will challenge the view that the behavior of scientists as they go about their work is less subject to the influence of social processes than is the behavior of people in other occupations. At the same time the book will call for caution in concluding, as do many contemporary sociologists of science, that the content of science is determined solely by social variables.

To study whether social factors influence the cognitive content of science, we must think carefully about what we mean by "cognitive content" and about what we want to understand. One way to characterize the content of science is by what Robert Merton has called its "foci of attention," or its selection of subject matter. How do scientists decide what research area to work in, and within a research area, how do they select particular problems to work on? There can be little doubt that the foci of attention of science are influenced by social variables. Indeed, Merton, in his famous monograph *Science, Technology, and Society in Seventeenth-Century England,* first published in 1938, showed how the problems that seventeenth-century British scientists selected were influenced by a series of economic and military concerns that confronted their society.[1] Merton argued that because of England's expanding overseas empires and increasing foreign trade, seventeenth-century English scientists focused their attention on scientific problems which would help improve ships' navigability. But he did not argue that the actual content of the theories that scientists developed while pursuing scientific problems was influenced by social variables. The concept of "foci of attention" allows us to make an analytic distinction between the social influences on the selection of what scientific problem to study and the scientific solution to that problem.

Another way to characterize the cognitive content of science is to look at the "rate of advance," or how quickly knowledge grows in a particular field. Merton's study of seventeenth-century English science also dealt with this problem, as well, in seeking to analyze the reasons for the rapid advance of science in the second half of the century. Merton saw the rate of advance as being influenced by the number of talented people who chose science as a career. In this part of his analysis he counted how many eminent people went into science as opposed to other areas such as the church or the military. Here again, however, he was not concerned with the actual content of the scientific papers or monographs published by the scientists in his study.

The third, and perhaps the most important, way to characterize the cognitive content of science is by the actual intellectual substance of scientific work. Thus some sociologists who study intellectual substance want to show how social variables and processes cause scientists to develop certain scientific ideas rather than others. If we can characterize the content of a particular scientific paper—"The Theory of Superconductivity," for example, the famous 1957 article by John Bardeen, Leon Cooper, and John Schrieffer—as x, then the sociology of science according to these practitioners should show how that paper came to be x rather than y or z. Most of this book is concerned with recent attempts to understand the content of science in this sense. That is, it examines attempts to show that science's intellectual content, as distinct from its foci of attention or rate of advance, is largely a function of social processes.

Until recently, most sociologists felt that the discipline could reveal very little about the content of the natural sciences. In the 1920s, the sociologist of knowledge Karl Mannheim ([1929] 1954) concluded that, although the content of social science ideas could be explained by examining the social contexts in which they were produced and the sociopolitical positions of their authors, the content of the natural sciences was not dependent upon these variables.[2] Following the then widely accepted positivist or realist view of science, he believed the content of the natural sciences to be ultimately determined by "facts" or "reality." In natural science, at least, a "truth" existed, and any scientific theory which did not express that truth would eventually be found to be inconsistent with empirical data and ultimately discarded. In this sense, natural scientists were trying to discover the next page of a book that had already been written, whose conclusion, though currently unknown, was predetermined or inevitable. Nature, rather than sociological processes, determined the way in which scientific knowledge developed.

This view, still held by the majority of scientists (including sociologists), was the underlying assumption of most work in the sociology of science until the 1970s. The belief that social factors were not a significant influence on the content of science was at least implicitly adopted by Merton, the founding father of the sociology of science. In *Science, Technology, and Society in Seventeenth-Century England* ([1938] 1970), he stated that he was not trying to explain the specific content of the scientific ideas developed. "Specific discoveries and inventions belong to the internal history of science and are largely independent of factors other than the purely scientific" (p. 75). In his philosophy of science, Merton, like the great majority of his contemporaries, was a positivist. "To a certain degree, a fixed order must prevail in the appearance of scientific discoveries; each discovery must await certain prerequisite developments" (p. 77). In the annotated version of his famous and now controversial 1942 paper on the "norms" of science reprinted in *Social Theory and Social Structure* (rev. ed. 1957a), Merton indicated his belief that sociology could not explain the content of science, which was ultimately determined by nature. Here Merton added a note in which he discussed the attempt by the leaders of the Soviet Union in the post–World War II period to define science as "national in form and class in content" (p. 554). He commented:

> This view confuses two distinct issues: first, the cultural context in any given nation or society may predispose scientists to focus on certain problems, to be sensitive to some and not other problems on the frontiers of science. This has long since been observed. But this is basically different from the second issue: the criteria of validity of claims to scientific knowledge are not matters of national taste and culture. *Sooner or later, competing claims to validity are settled by the universalistic facts of nature which are consonant with one and not with another theory.* (p. 554, italics added)

When this essay was reprinted in the 1973 collection of Merton's papers on the sociology of science, Merton altered the last sentence to read: "Sooner or later, competing claims to validity are settled by universalistic criteria" (p. 271).

In a 1971 book, Joseph Ben-David, another early leader in the sociology of science, concluded that "the possibilities for either an interactional or institutional sociology of the conceptual and theoretical contents of science are extremely limited" (pp. 13–14). Up until the 1970s sociologists of science, including Paul Allison, Bernard Barber, Ben-David, Cole and Cole, Diana Crane, Jerry Gaston, Warren Hagstrom, Lowell Hargens, Norman Kaplan, Merton, Norman Storer, and Harriet Zuckerman, studied the internal social organization of the scientific community and paid very little attention to the cognitive content of science. The papers and books produced by scientists remained, in Richard Whitley's (1972) term, a "black box."

This situation changed rapidly in the 1970s. Influenced by the revolution which had occurred in the history and philosophy of science, sociologists concentrated their attention on explaining the content of ideas.[3] Although both American and European sociologists studied this problem, the work of the latter became increasingly important. The approach to the sociology of science dominant today was first called "relativism-constructivism" and is now commonly referred to as "social constructivism." Its leading practitioners include Barry Barnes, David Bloor, Harry M. Collins, Michael Mulkay, and Steve Woolgar in Great Britain; Bruno Latour in France; and Karin Knorr-Cetina in Germany.

The social constructivists base their research on several major assumptions. First, they believe that science is not a rule-governed activity; scientists do not follow a set of procedures which enable independent workers to agree upon what is true. Second, they believe that scientific disputes frequently cannot be resolved by empirical evidence. Because evidence is only meaningful in the light of the theory which led to its production, evidence cannot settle disputes among competing theories. Third, and most important, they adopt a relativist philosophical position and deny the importance of nature as an objective external which influences the content of scientific knowledge. Nature does not determine science; instead, they say, the social behavior of the scientists in the laboratory determines how the laws of nature are defined. Given these assumptions, the social behavior of the scientists becomes the most important determinant of the content of their ideas. Thus it is possible to show how social variables influenced "The Theory of Superconductivity" or to show how any other cognitive work takes the form it does rather than some other possible form.

In this book I report the results of theoretical and empirical investigations I have undertaken related to the question of how the cognitive aspects of science are influenced by social variables. The position I take will be contrasted with three others. The first is the position of traditional positivist historians and philosophers of science, who include philosophers such as Rudolf Carnap and Ernest Nagel and historians such as Charles C. Gillispie, A. Rupert Hall, and George Sarton. Positivism was the dominant view in the history and philosophy of science until the 1960s. It is still the view held by most nonscientists and almost all working scientists, including sociologists. I treat positivism here as an ideal type, and set aside the many subtleties and variations in the approach of the positivist philosophers. Second is the position of those whom I call the new historians and philosophers of science. Their views are best illustrated by Thomas Kuhn's classic book, *The Structure of Scientific Revolutions* ([1962] 1970). Kuhn, Imre Lakatos (1970), Paul Feyerabend (1975), and Michael Polanyi (1958) challenged many of the assumptions of the positivists. To some extent I am also treating this group's work as an ideal type, for the purpose of contrasting the traditional with the more modern approach to science. Third is the position of the social constructivists, briefly outlined above. The differences among the three groups will become clearer as we consider how each views the important question of consensus in science, or how agreement is reached on what new scientific work should be accepted as important and true.

The Positivists' View of Consensus in Science

The positivists had a clear explanation for the problem of consensus. They saw science as differing from other areas of human endeavor in the ability of scientists to achieve consensus based upon the dispassionate evaluation of objective empirical evidence. In this view the establishment of consensus was nonproblematic; new contributions to science which were true and could be shown to fit the facts of nature were the ones which came to be accepted by the scientific community. Nature and empirical evidence were seen as the primary and ultimate arbiters of scientific dispute and the enforcers of consensus. According to positivist philosophy of science, the objective validity of a scientific contribution could be determined by using a set of rules to evaluate evidence. Laudan (1984) refers to this belief as the "Leibnizian ideal":

> At least since Bacon, most philosophers have believed there to be an algorithm or set of algorithms which would permit any impartial observer to judge the degree to which a certain body of data rendered different explanations of those data true or false, probable or improbable . . . Science was regarded as consisting entirely in claims about matters of fact. Since scientific disagreements were thought to be, at bottom, disagreements about matters of fact, and since disagreements of that sort were thought to be mechanically resolvable, philosophers had a ready sketch of an explanation for consensus formation in science. (pp. 5–6)

The reaction of the scientific community to newly published work was believed to be determined by the content of that work. Those papers which contained empirical evidence which could be replicated by others and those theories or models which were supported by empirical evidence would be accepted as valid; those papers whose content could not be replicated would be rejected. Positivists would also argue that the current cognitive state of a discipline would make some cognitive problems inherently more important than others.

New papers which successfully dealt with these problems would be the most likely to be accepted as facts. Further, the content of what could be accepted as fact was determined by nature. Scientists were trying to discover the laws of nature, which was a given and an external over which they had no control. If a paper put forth an idea which was not in accord with the laws of nature, attempts to replicate or support this idea would ultimately be doomed. Positivists sometimes saw social factors as influencing the reception of new ideas, but only as impediments. Barber (1961) shows how a scientist's position in the scientific community is one factor that might cause resistance to a correct idea, as in the case of Mendel.

Kuhn's View of Consensus in Science

Kuhn divides science up into two phases: normal science and revolutionary science. The level of consensus during normal science is high; virtually all scientists working in a field accept the truth and legitimacy of the current "paradigm."[4] During the rare periods of revolutionary science, such as the transformation from classical mechanics to quantum mechanics in physics, consensus on the paradigm will break down. But in a relatively short period of time a new consensus will emerge around the new paradigm.

But how are new ideas evaluated? Here there is a sharp difference between the view of Kuhn and that of the positivists. Following philosophers such as Pierre Duhem and W. V. Quine, Kuhn rejects the notion that empirical observations are objective; different scientists can formulate differing and sometimes diametrically opposed interpretations of the same data. When he looks at the relationship between theory and observations, he finds that the observations are influenced at least as much by theory as the other way around. Data are thus theory-laden. The views of science projected by the writings of Kuhn give primacy in science to theory. Rather than experimental and observational data being the determinant of the course of science, theories determine what evidence is looked for and what evidence is taken seriously.

In a paper on the functions of measurement in the modern physical sciences Kuhn stresses that the real function of experiment is not the testing of theories. Theories are frequently accepted before there is empirical evidence to support them. Results which confirm already accepted theories are paid attention to, while disconfirming results are ignored. Kuhn cites Dalton's law of multiple proportions and the research done on it:

> Needless to say, Dalton's search of the literature yielded some data that, in his view, sufficiently supported the Law. But—and this is the point of the illustration—much of the then extant data did *not* support Dalton's Law at all . . . But, at the beginning of the nineteenth century, chemists did not know how to perform quantitative analyses that displayed multiple proportions. By 1850 they had learned, *but only by letting Dalton's theory lead them.* Knowing what results they should expect from chemical analyses, chemists were able to devise techniques that got them. As a result chemistry texts can now state that quantitative analysis confirms Dalton's atomism and forget that, historically, the relevant analytic techniques are based upon the very theory they are said to confirm. *Before Dalton's theory was announced, measurement did not give the same results.* There are self-fulfilling prophecies in the physical as well as in the social sciences. (Kuhn [1961] 1977, pp. 195–196, italics added)

This example is one illustration of the way in which positivist historical reconstructions have made it appear as if the acceptance of theory by the scientific community was dependent on empirical supporting evidence.[5] But according to Kuhn, during normal science scientists accept almost without question (or, as Karl Popper calls it, without "criticism") the dominant scientific theories in their research area, even if there is empirical evidence which the theory is unable to explain or which suggests that the theory is wrong. Failure of an experiment to confirm a theory or to achieve results expected by a widely accepted theory casts doubt not upon the theory but upon the ability of the experimenter.

Kuhn ([1962] 1970) gives several examples of theories maintained in the face of significant counterevidence or anomalies. Even the most important theories cannot explain all the known facts: "During the sixty years after Newton's original computation, the predicted motion of the moon's perigee remained only half of that observed. As Europe's best mathematical physicists continued to wrestle unsuccessfully with the well-known discrepancy, there were occasional proposals for a modification of Newton's inverse square law. But no one took these proposals very seriously, and in practice this patience with a major anomaly proved justified" (p. 81).

An issue related to the significance of empirical observations is the more philosophical question of the extent to which science can be said to be characterized by progress. Kuhn argued that the traditional portrayal of the history of science was wrong and represented a severe distortion of what actually happens in science. Positivists saw science as progressing because, at each moment in time, current theories are a better approximation to nature's laws than past theories. According to this view, it is unnecessary to conclude that Newton was wrong if you accept Einstein as being right. Newtonian theory was merely a special case of the new theory of relativity.

Kuhn decisively rejects the positivists' view of science. Although he does not deny that a theory at Time 2 is "better" than a theory at Time 1, he does not use approximation to "truth" as a criterion in evaluating theories or paradigms. In fact, the concept of objective truth plays no role in Kuhn's analysis. For Kuhn a new paradigm is simply a different, rather than a better, way of looking at reality. He points out that when a new paradigm is adopted, it is sometimes inferior to the one it replaces in its ability to explain a wide range of empirical phenomena. In rejecting one paradigm for a new one "there are losses as well as gains" ([1962] 1970, p. 167). By this he means that there are some phenomena which will never be explained as well by the new paradigm as by the old and that there are some problems which can be more easily solved using the old paradigm.[6] Engineers find the assumptions of Newtonian mechanics more useful than those of quantum mechanics in solving some problems, for example.

Kuhn rejects the notion that science is cumulative in the sense that a new theory is necessarily an improvement on the one it replaces. He finds that science changes not in an evolutionary fashion but in a series of revolutions in which the world view of the members of the scientific community is drastically altered. It is important to emphasize that Kuhn sees competing paradigms as incommensurable. "Scientific revolutions are here taken to be those non-cumulative developmental episodes in which an older paradigm is replaced in whole or in part by an incompatible new one" ([1962] 1970, p. 92).[7] He thus concludes that "Einstein's theory can be accepted only with the recognition that Newton's was wrong" (p. 98).

This view of the history of science not only contradicts most previous interpretations but is in sharp contrast with the view held by and still taught to working scientists. It is, for example, difficult to find a practicing physicist who will agree with Kuhn's conclusion that if Einstein is right then Newton had to be wrong. According to Kuhn, the differences between the new view of science and that held by working scientists can be explained by differences in the source of knowledge.

Physicists learn about the theories of Newton and Einstein from text-books. But whenever a revolution occurs in science, the textbooks are rewritten and the past history of the field is reconstructed in a way that makes prior paradigms appear to lead to the current paradigm.

Kuhn argues that it is only by distorting the work of earlier scientists, sometimes until it is hardly recognizable as the same work, that earlier work can be made to seem to be encompassed by a later, incommensurable paradigm. Essentially the textbook writers and the older historians of science have decontextualized the original science and reinterpreted it in light of current knowledge. The reinterpretation would be impossible were it not for the current knowledge.

Kuhn rejects the assumption that one idea can be proved to be better than another by empirical data. He argues that when a new paradigm does emerge, it is not adopted on empirical grounds. He concludes that "the competition between paradigms is not the sort of battle that can be resolved by proofs" ([1962] 1970, p. 148). Kuhn's position leaves the question of consensus unanswered. If ideas are not tested with objective empirical observations and if the choice between two competing theories cannot be made on the basis of fit with evidence, how is this choice made? Kuhn does not provide a detailed answer to this question. Sometimes he implies that scientists will make their choices based upon aesthetic criteria or upon which paradigm will yield the most new puzzles. But he never gives a full description of how scientists actually use such criteria.

The Constructivists' View of Consensus in Science

Constructivist sociologists of science see Kuhn's position as a warrant for adopting a relativist epistemology. Because relativist philosophy of science plays such an important role in recent sociology of science, it is important that the distinction between the "realist" and "relativist" positions be clear. In his article "Two-Tier-Thinking: Philosophical Realism and Historical Relativism," the historian of science Yehuda Elkana (1978) discusses these two philosophical traditions. He defines the realist position:

> What all this boils down to is that experimental evidence determines fairly well the structure of scientific theories and we have the means to order theories according to their truth value; some theories are true, some others are false—and among those that are true there is a hierar-

chical order on which theories can be held to be more or less "experimentally proved." (p. 311)

And the relativist position:

> Cognitive relativism is the view that truth and logic are always formulated in the framework of, and are relative to, a given thought-world with its own language. Before it all became fashionable, Wittgenstein used to say, "All testing and confirmation and disconfirmation of a hypothesis takes place already within a system": and Quine, "Where it makes sense to apply 'truth' is to a sentence couched in the terms of a given theory and seen from within this theory, complete with its posited reality." A beautiful formulation of relativism is that of Mary Douglas: "It is no more easy to defend non-context dependent, non-culture dependent beliefs of things or objective scientific truths than beliefs in gods and demons." (p. 312)

Thus relativists would argue that when the Hopi Indians engage in their ritual rain dance their interpretation of the dance as causing rain is just as justifiable as the meteorologists' argument that the rain dance cannot make it rain. Both are dependent upon contradictory assumptions, neither set of which can be proved to be right. Relativists argue that no group is more logical or rational than another (Latour, 1987, p. 195).

Clearly scientists working at the research frontier do not know what the laws of nature are, and that well-trained scientists can reach very different solutions in their attempt to "read" those laws illustrates that the content of new knowledge is "underdetermined." Essentially, for any scientific problem, the existing empirical data leave substantial room for divergent views on what the solution to the problem is. It is also clear that much of what was commonly accepted by the scientific community as true in the past is currently believed to be wrong. What we currently believe to be true will in the future most likely be thought of as wrong. That a statement is accepted by scientists as a fact and is therefore seen by scientists as determined by nature does not make this statement true, Latour (1987) argues. Using James Watson's account of the discovery of the double helix, Latour describes how Watson relied upon information contained in chemistry textbooks. But a chemist who shared an office with Watson, Jerry Donohue, told him that the textbooks were wrong, "that for years organic chemists had been arbitrarily favoring particular tautomeric forms over their alternatives on only the flimsiest of grounds" (Watson, 1968, quoted in Latour, 1987, p. 8). Donohue turned out to be correct and all the textbooks incorrect. Given that facts can easily become errors, what sense does it make to see what is at Time 1 a "fact" and at Time 2 an "error" as being *determined* by nature?

The position of the constructivists is that the content of solutions to scientific problems is developed in a social context and through a series of social processes. In this sense the content of science is socially constructed. Each new discovery is a result not of the application of a set of rational rules for evaluating empirical evidence but of chance occurrences, the availability of particular equipment, the availability by chance of particular substances, and social negotiation among people inside the laboratory and sometimes between those inside the laboratory and those outside the laboratory. Only in writing up the discovery for an audience of colleagues do the scientists fit what they have done into a "story" emphasizing rationality and expunging all the chance circumstances and social negotiation which led to the discovery (Knorr-Cetina, 1981).

The traditional view of science is supported by a type of history in which one looks back from the present and reinterprets all that came before as steps in an inevitable process. The constructivist view of science is supported by a history in which one goes back in time and then looks forward. When this is done it can easily be seen that what is accepted as true at any given point in time can and usually does come to be rejected as wrong sometime in the future. This approach to the history of science raises serious doubt about whether there are any external laws of nature which influence what is held to be true at any given point in time.

Most of those sociologists who call themselves relativists today, however, go further than denying that nature determines what is believed to be true. They argue that the empirical world has little, if any, influence on what is accepted as true by the scientific community. For example, Collins (1981, p. 3) concludes that the "natural world has a small or non-existent role in the construction of scientific knowledge."[8]

The constructivists back up their epistemological stand with a series of micro-level studies of science which lead to the conclusion that scientists do not behave in the ways that the positivist philosophers said that they should. Consider, for example, the role of the replication of experiments in science. Traditionally, as Zuckerman (1977b) points out, replication has been considered the basis of social control in science and a means of preventing fraud. Because scientists know that their work can and will be replicated, they are motivated to do the work carefully and are inhibited from publishing sloppy or outright fraudulent results. And, most important, if science is indeed based upon replication, the idiosyncratic factors which influence the doing of science in any particular context will not have a major influence on the development of knowledge in a research area. Replication has been thought to make science rational in the sense that consensus is based upon empirical evidence obtained from experiments which can be reproduced by any competent member of the scientific community.

The problem with this position, in the view of Harry M. Collins, is that replication is almost never done. In his article "The Seven Sexes: A Study in the Sociology of a Phenomenon, or the Replication of Experiments in Physics," Collins (1975) analyzes the work of physicists studying gravity waves. He finds that scientists have very little interest in conducting replications of other scientists' work: "Workers in the gravity wave field explained their lack of interest in producing carbon copies of the originator's apparatus in several ways. Some said that an exact copy could gain them no prestige. If it confirmed the first researcher's findings, it would do nothing for *them,* but would win a Nobel prize for *him,* while on the other hand, if it disconfirmed the results there would be nothing positive to show for their work" (p. 210).[9]

The actual doing of experiments requires what Polanyi (1958) has called "tacit knowledge," idiosyncratic knowledge required to do experiments that can only be learned by being a participant. Most journals refuse to publish extended discussions of technique; they are considered not "science" but rather an aspect of the technology of doing science. The techniques of doing research in most research areas are highly idiosyncratic. The selection of equipment for use in experiments is based on what is available in the laboratory at the time or the particular knowledge or skills of technicians. Frequently parameters used in experiments are selected by sheer chance. The temperature of water used in a biological sciences experiment, for example, might be determined simply by the temperature of the water that came out of the tap in the laboratory.[10] Collins (1974) argues that given the highly idiosyncratic aspects of devising equipment, getting the equipment to work, and setting parameters in experiments, all of which are usually not reported in journal articles, it is frequently impossible to replicate an experiment by reading the journal article.

Constructivists argue that the cognitive content of new papers has no influence on how they are received. Reception is influenced by a set of rhetorical strategies used by the paper's author to convince others to accept the paper as true. Allies are enlisted through the process of citation even if the citation distorts the intent of the author being cited (Latour, 1987). The reception of the new paper will also be influenced by the social characteristics of its author and the operation of social processes such as intellectual authority. Those who evaluate the papers have a distinct set of both cognitive and social interests, and these interests, rather than any characteristic of the work being evaluated, determine what work they choose to accept. Consensus is attained when a new discovery becomes so powerful through the allies it has successfully enlisted that the effort required to overturn it is too great.

Consensus in Science

The position I take in this book is in agreement with that of the constructivists on some points and in sharp disagreement on others. Recent research has convinced me, as it has virtually all other members of the discipline which has come to be called the "social study of science," that science is not practiced in the rational rule-governed way described in the positivist philosophy of science texts. Rather than being a highly objective enterprise based upon the empirical verification of theories, science is a much less objective enterprise in which theory holds at least an equal role to evidence and in which consensus is fragile and changing.[11] Woolgar (1988, p. 107) writes: "Perhaps the most significant achievement of the social study of science is the finding that the natural sciences themselves only rarely live up to the ideals of SCIENCE!" On this point, then, there is agreement between Kuhn, the constructivists, and myself; we each take a position differing from that of the positivists.

On the crucial issue of consensus the constructivists and I are in agreement on the level of consensus observed in the natural sciences but disagree on the processes through which consensus is achieved. Let us first consider the extent to which the natural sciences are characterized by cognitive consensus. Here it is necessary to introduce an important conceptual distinction between two types of knowledge: the *core* and the *research frontier*.

The Core and the Frontier

The core consists of a small set of theories, analytic techniques, and facts which represent the given at any particular point in time. If we were to look at the content of courses taught advanced undergraduate students or first-year graduate students in a field such as physics, we would acquire a good idea of the content of physics' core of knowledge. The core is the starting point, the knowledge which people take as a given from which new knowledge will be produced. Generally the core of a research area contains a relatively small number of theories, on which there is substantial consensus. The core consists of what Latour and Woolgar ([1979] 1986) call "facts": knowledge which has been accepted by the scientific community as true or as an adequate representation of nature. In addition, core knowledge must be judged by the community to be "important." A large part of research frontier science works at producing low-level descriptive analyses rather than at solving any particular problem. In order for a new scientific contribution to get into the core it must be judged to be both true and important. Most new scientific contributions never have a chance to enter the core because they fail to meet the latter criterion. For this type of science few members of the community will even be concerned with whether it is true or not, and therefore underdetermination will not be an important factor influencing evaluation.

The other component of knowledge, the research frontier, consists of all the work currently being produced by all active researchers in a given discipline. The research frontier is where all new knowledge is produced. Most observers of science agree that the crucial variable differentiating core and frontier knowledge is the presence or absence of consensus. A particular contribution to knowledge does not become part of the core until a substantial consensus on it is established. The physicist and sociologist of science John Ziman describes what happens to physics as it passes from the frontier to the core: "At the 'frontiers,' where scientific knowledge is being laboriously acquired, *Physics is as crude and uncertain as any other discipline,* but it has the power to subdue, chain down, discipline and civilize new theories, once they are reasonably established, so that they become as 'absolute,' within a few years, as the most venerable principles of Newton or Faraday or Maxwell" (1968, p. 42, italics added).

Later Ziman compares the core physics described in undergraduate texts with the frontier knowledge that the advanced graduate student confronts: "The scientific literature is strewn with half-finished work, more or less correct but not completed with such care and generality as to settle the matter once and for all. The tidy comprehensiveness of undergraduate Science, marshalled by the brisk pens of the latest complacent generation of textbook writers, gives way to a nondescript land, of bits and pieces and yawning gaps, vast fruitless edifices and tiny elegant masterpieces, through which the graduate student is expected to find his way with only a muddled review article as a guide" (1968, p. 73).

The research frontier is linked to the core through the evaluation process. A large majority of new scientific contributions produced at the frontier turn out to be of little or no lasting significance (J. R. Cole and S. Cole, 1972). For example, consider a sample of 1,187 papers published in 1963 in the *Physical Review,* one of the two most prestigious physics journals in the world. Three years after publication about half of these papers had received one or no citations; only 15 percent had received six or more citations (S. Cole, 1970). Most of the work produced at the frontier has little or no impact on the development of community-based knowledge.

It is necessary for sociologists of science to make this distinction between core and frontier because the social character of knowledge in these two components differs dramatically. Core knowledge is characterized by virtually universal consensus. Scientists accept this knowledge as a given and as a starting point for their research. Empirical evidence which might not fit in with core knowledge is generally overlooked or rejected. Because scientists accept core knowledge as being true, they see its content as determined by the laws of nature. In frontier knowledge different scientists looking at the same empirical evidence can reach different conclusions. Frontier knowledge is accepted by scientists not as true but as the *claims* to truth of particular scientists. If we look only at core knowledge and at what scientists say about core knowledge, we will conclude that science is adequately described by the traditional view. If we look at frontier knowledge, however, we will find little confirmation for much of the traditional view. Determining how much consensus there is in a particular scientific research area is a difficult problem, both methodologically and conceptually, and we shall return to it in a later section.

Limited Obscurantism

There can be significant variation in consensus even about parts of the scientific core. Although by definition there is substantial consensus that certain theories, methods, and exemplars belong to the core, when we ask what the consensus consists of we may find that there is consensus only on the importance of a theory or on its most general characteristics and significant disagreement about application and details.

Kuhn ([1962] 1970) pointed out that scientists frequently agree in identifying a paradigm that they all accept, but may not fully agree on the interpretation of that paradigm: "Scientists can agree that a Newton, Lavoisier, Maxwell, or Einstein has produced an apparently permanent solution to a group of outstanding problems and still disagree, sometimes without being aware of it, about the particular abstract characteristics that make those solutions permanent. They can, that is, agree in their *identification* of a paradigm without agreeing on, or even attempting to produce, a full *interpretation* or *rationalization* of it" (p. 44).

In a 1990 episode of the Public Broadcasting System show *Nova* focusing on quantum theory, one physicist reported that if you asked ten different physicists to explain quantum theory you would get nine different answers. In the same vein, Gilbert and Mulkay (1984) interviewed scientists who had worked on the chemiosmotic theory, for which Peter Mitchell was awarded the Nobel Prize in chemistry. Gilbert and Mulkay found that accounts of the theory given by the scientists varied so much that it was doubtful whether there could be said to be any consensus on it:

> The interpretative variability of "chemiosmosis" is not easily discerned in the ordinary course of events. Much of the time, it is hidden by the character of scientists' discourse about consensus. For researchers regularly speak as if "chemiosmosis" is an entity held in common with most other colleagues. They each proceed as if the specific version of the theory that they are engaged in proposing is "the real chemiosmosis" which is coming to be accepted or rejected by the field. They continually construct their accounts as if they are referring to "a theory" which

exists independently of their interpretative work. However, the detailed comparisons between accounts carried out above reveal that the apparent facticity of chemiosmosis and its apparently widespread endorsement are illusory in the specific sense that they exist, not as objective entities in an external social world, but only as attributes of participants' contingent consensus accounts. (pp. 136–137)

Because individual participants express differing views of what the consensus is or should be, Gilbert and Mulkay come to the conclusion that there can be no meaningful analysis of consensus. But their own data suggest that all the scientists interviewed would agree that the chemiosmotic theory was or had been an important one. There is a vague general version of the theory that all agree upon. It is upon the details of the theory that there remains disagreement. This example is useful in bringing to our attention that even in parts of the core there can be considerable disagreement on interpretations and understanding. Science seems to tolerate high levels of ambiguity and disagreement and in some areas of science it is possible that ambiguity in a theory will be a characteristic which will contribute to the theory's perceived importance. Ambiguous but "interesting" theories will allow many different scientists to use them in varying ways. The ambiguous theory becomes a stimulus for additional research and permits substantial individual creativity while maintaining the consensus necessary for scientific work to proceed. This role might be referred to as the function of "limited obscurantism."

Think of expressing a theory in three possible ways. In version A the theory could be stated so clearly that almost all members of an audience would interpret it in the same way. In version B the theory could be stated in such an obscure way that most members of an audience would have difficulty interpreting it at all. In version C, the version making use of "limited obscurantism," the theory could be stated in a slightly obscure way, so that various members of an audience would interpret it differently, depending upon their own intellectual backgrounds and interests. If we were to present these three different versions of the theory to an audience which had been randomly divided into three groups, we might find the group exposed to version C to have a more favorable opinion of the theory than those exposed to versions A and B. Version A might be seen as too "obvious" and as yielding little new information. Version B would elicit unfavorable response because it could not be understood. Version C, with its ability to appeal to different interests, could very well be the version eliciting the most favorable opinion.

There are many examples in the social sciences of theories, including Kuhn's theory of scientific revolutions, which have benefited substantially from limited obscurantism. Tolerance for ambiguity is necessary if the work of science is to proceed.[12] Although we readily recognize this as true for the social sciences, the example given by Gilbert and Mulkay and many other examples in the literature demonstrate that a similar situation exists in the natural sciences as well.

Level of Consensus at the Frontier

By definition there is a high level of agreement on core knowledge, but what about research frontier knowledge? An examination of frontier knowledge as opposed to core knowledge supports the view that science is not a rule-governed activity which enables scientists to achieve consensus. In Chapter 4 I will present the results of empirical studies which demonstrate that the level of cognitive consensus at the frontier is relatively low in all scientific fields.[13] These studies examined the extent to which reviewers of proposals to the National Science Foundation (NSF) agreed in their evaluations. In all fields studied, natural sciences as well as social sciences, qualified reviewers frequently reached different conclusions in evaluating the proposals. The disagreement was so prevalent that approval of a proposal submitted to the NSF was due as much to "luck" as to the quality of the proposal. Because different qualified reviewers randomly selected from the pool of eligible reviewers will evaluate the same proposal differently, the acceptance of a proposal depends to a great extent on luck in having one's proposal sent to reviewers who will look upon it favorably. This study leads to the conclusion that there may not be significantly more consensus in evaluating new scientific ideas than there is in judging nonscientific items such as human beauty, new works of art, or Bordeaux wines.

Whether or not there are any rules which scientists follow in doing their work, these rules do not lead to high levels of consensus. Positivists have claimed that science differs from other types of intellectual pursuit in that science displays high levels of consensus. I conclude from my studies of consensus at the research frontier that for the large majority of new scientific work the cognitive content will not determine its reception.

Kuhn saw science as displaying high levels of consensus once a paradigm had been adopted by a scientific community, but experiencing low levels of consensus during the rare but important paradigm debates. Kuhn, however, does not adequately distinguish between the core and the frontier. Kuhn emphasizes the former, on which there is consensus, and deemphasizes the latter, on which there is often little consensus. It is my position that in considering the question of consensus we cannot refer to "science" as a uniform whole but must consider the research frontier and the core separately. I differ from Kuhn in finding much lower levels of consensus among scientists as they go about their everyday work.

The social constructivists and I agree on the level of consensus at the research frontier. Some constructivists such as Latour and Woolgar ([1979] 1986) make a distinction between levels of "facticity" which is similar to the differentiation I make between the core and the frontier. In addition, I find that there are no significant differences in the level of consensus at the frontier between the natural and the social sciences, a position supported by most constructivists (see Chapter 5) but rejected by positivists, Kuhn, and most traditional sociologists of science. The macro-level work I have done on consensus fits in with and supports the view of the constructivists that science is underdetermined by empirical evidence and rules of procedure. Both the micro-level research of the constructivists and my macro-level research on consensus in various scientific fields support the new view of science which casts doubt on the existence of a set of rational criteria which enable scientists to reach similar conclusions.

How Consensus Is Achieved

Although the social constructivists and I agree on the extent to which science is characterized by consensus, there is substantial disagreement between us on the issue of how core consensus emerges from the seeming chaos of frontier science. I will argue that the reception of new scientific work is influenced by three sets of interacting variables: the content of the work itself, the social characteristics of the authors, and the operation of social processes such as intellectual authority. Consider a newly published scientific paper. What are the possible variables which could influence its reception by the scientific community? It is clear that the characteristics of the author could influence its reception. The new work of eminent scientists may be more likely to receive attention than the new work of unknown scientists. This hypothesis will be discussed below and more extensively in Chapters 6 and 7.

The reaction of individual scientists to a new paper could also be influenced by the reaction of other members of the scientific community. Because our views of what is good or bad and right or wrong are strongly influenced by the opinions of intellectual authorities, the reception of a new paper can be significantly affected by their reaction. This process will be discussed below and more extensively in Chapter 8.

The constructivists and I agree on the importance of both author characteristics and social processes as influences on the formation of consensus. But to what extent will the reception of a new paper be influenced simply by the content of the empirical evidence, theories, and models presented in the paper? I believe that the cognitive content of a paper can sometimes have a significant independent influence on its reception; the constructivists argue that it has no influence.

My position is that although the content of most papers does not compel acceptance, the content of a few will. Consider doing an experiment in which a group of authorities would be asked to evaluate papers prior to their publication without knowing the names of their authors. Suppose that prior to its publication a group of experts had been asked their opinion of an anonymous draft of the 1953 paper in which James Watson and Francis Crick announced their model of DNA. Would there have been substantial consensus that this work was both important and true? I say that there would. The constructivists would have to say that there would not, because if there were substantial consensus it would be an example of how content alone can sometimes compel acceptance. Some constructivists might try to get out of this bind by denying even an analytic distinction between the content of a paper and social processes, of course. But as I will discuss in Chapter 3, this approach leads to an unsatisfactory tautology and ignores the most interesting questions which concern the interaction of cognitive and social variables in influencing evaluation of new scientific contributions.

The most important evidence in support of my position is the fact that some discoveries are almost immediately accepted by the scientific community as being true and are quickly added to the core.

Among the relatively rare contributions that do have some impact on the development of community-based knowledge, the great majority are recognized as important immediately or soon after publication. This was the conclusion of my earlier study of "resistance" to scientific discovery or "delayed recognition" (S. Cole, 1970). I conclude that the evaluation system of contemporary science operates so that only a relatively small number of papers that later turn out to be significant are overlooked at the time of their publication. This does not mean, of course, that they are immediately accepted as part of the core; there is some variation in the time it takes for a new contribution to enter the core. A contribution can be recognized as important but remain controversial. Thus some new knowledge, such as the discovery by Watson and Crick of the structure of DNA, turns out to be almost immediately accepted as part of the core.[14] Other important work, such as that described by Gilbert and Mulkay (1984) on the chemiosmotic theory, takes many years before it becomes fully accepted and is added to the core. The three examples of discoveries that met with quick and almost universal acceptance that will be used here are the Watson and Crick model of DNA, the Bardeen, Cooper, and Schrieffer theory of superconductivity, and the discovery by R. Guillemin and A. V. Schally of Thyrotropin Releasing Factor (TRF). The rapid acceptance of these discoveries suggests that there was some aspect of their content which influenced other scientists to accept them as true.

Some constructivists might agree with this position but counter that the attributes which compel acceptance have been socially constructed in the past. This argument raises the deeper and more philosophical issue of the extent to which solutions to scientific problems in either the past or the present are constrained by empirical evidence from the natural world. The constructivists argue that although the scientific community may be constrained by past social constructions, these have nothing to do with constraint by the natural world. My position is that both past social constructions and current research are to at least some significant extent constrained by evidence from the natural world. The constructivists have misinterpreted Kuhn's position as suggesting that empirical evidence does not constrain the cognitive content of science. *Sometimes* it may be impossible to use evidence from the external empirical world to determine which of two competing scientific ideas should be accepted. According to Kuhn, this is usually the case in paradigm debates. But even here it would be possible to misinterpret Kuhn's position and underemphasize the importance of empirical observation. That it is impossible to use empirical evidence to choose between two specific competitors in a paradigm debate does *not* mean that it would be impossible to use such evidence to choose among any competitors.

In his paper on measurement, Kuhn ([1961] 1977), discussing the importance of theory in determining what data scientists look for and how they interpret those data, states that this does not mean that any theory will be accepted: "If what I have said is right, nature undoubtedly responds to the theoretical predispositions with which she is approached by the measuring scientist. But that is not to say either that nature will respond to any theory at all or that she will ever respond very much" (p. 200).

Kuhn goes on to discuss the caloric and dynamical theories of heat. These were two very different theories, but because the predictions from each were similar, any data generated would equally support both theories. Kuhn argues, however, that the data's support of both of these theories does not mean that the data would have supported any theory:

> There are logically possible theories of, say, heat that no sane scientist could ever have made nature fit . . . But those . . . merely "conceivable" theories are not among the options open to the practicing scientist. His concern is with theories that seem to fit what is known about nature, and all these theories, however different their structure, will necessarily seem to yield very similar predictive results. If they can be distinguished at all by measurements, those measurements will usually strain the limits of existing experimental techniques. Furthermore, within the limits imposed by those techniques, the numerical differences at issue will very often prove to be quite small. Only under these conditions and within these limits can one expect nature to respond to preconception. (p. 201)

Laudan (1984) expresses a similar position. He argues that although there may be many cases in which it will be impossible to utilize a set of rational criteria to choose between two theories, this does not mean that there are no rules which allow scientists to make rational choices among alternative scientific explanations:

> The crucial point here is that even when a rule underdetermines choice in the abstract, that same rule may still unambiguously dictate a comparative preference among extant alternatives . . . For instance, the rules and evidence of biology, although they do not establish the unique correctness of evolutionary theory, do exclude numerous creationist

hypotheses—for example, the claim that the earth is between 10,000 and 20,000 years old—from the permissible realm and thus provide a warrant for a rational preference for evolutionary over creationist biology . . . the slide from underdetermination to what we might call cognitive egalitarianism (i.e., the thesis that all beliefs are epistemically or evidentially equal in terms of their support) must be resisted, for it confusedly takes the fact that our rules fail to pick out a unique theory that satisfied them as a warrant for asserting the inability of the rules to make any discriminations whatever with respect to which theories are or are not in compliance with them. (pp. 29–30)

Although there is indeed much historical evidence to reject the traditional positivist view of the relationship between empirical evidence and theory, there is also much evidence to suggest that many constructivists have gone too far in reducing the importance of empirical evidence in the development of scientific knowledge. Contemporary historians such as Galison (1987) and Rudwick (1985) and philosophers such as Giere (1988) have conducted detailed case studies which emphasize the role of empirical evidence in settling scientific controversies. Galison's book, *How Experiments End,* stands out for its careful analysis of the interplay between theory and research and its argument against some of the extreme views taken by constructivists. Although it would be incorrect to see nature as presenting empirical evidence which always acts as the arbiter of consensus, it would be equally incorrect to conclude that evidence from the external world never places important limits on what the scientific community is likely to accept as true. Because science is underdetermined does not mean that it is completely undetermined.

The Epistemological Question

If science were as underdetermined as the relativists would have us believe, then it would be necessary to accept the conclusion that a scientific problem has many different solutions which could succeed in being accepted as true and which would allow the further growth of the research area. The social constructivists are forced to argue that, rather than the Watson and Crick model of DNA, some different model could have been proposed, been quickly accepted as truth, and been equally important in the explosion of research which followed the Watson and Crick discovery. Although I agree that over a very long period of time all scientific interpretations can come to be viewed as wrong, I believe that in a shorter time frame and within the boundaries of currently accepted knowledge there are some ideas or problem solutions which are right or which work better than others. I also believe that data generated from the empirical world play an important role in the decisions of the scientific community about the acceptability of particular solutions. That the Watson and Crick model is likely to be "ultimately" seen as "wrong," as is all currently accepted scientific knowledge, does not mean *that at the time it was proposed* that any other theory could have come to be accepted as "right."

I reject the relativism of the social constructivists, not on philosophical or historical grounds, but on sociological grounds. I agree with the constructivists that when one takes a long time frame and the broadest view of science, it may be impossible to determine whether one scientific theory represents a closer approximation to the laws of nature than another. But within the confines of a paradigm or a set of theoretical assumptions, it sometimes is possible to select from among two competing contributions on the basis of the fit between theories, models, and data in the contribution and evidence from the empirical world. My analysis is primarily concerned not with the relatively rare sweeping changes in paradigms, such as the switch from classical mechanics to quantum mechanics, but with the every-day doing of science. On a day-to-day basis virtually all scientists, even Nobel Prize winners, are working within the confines of a paradigm and doing "normal" science. Even though in some "ultimate" sense there may be no way to determine whether one paradigm is a better approximation to the "real" laws of nature than another, the exclusion of nature and the empirical world from our model of how scientific knowledge grows makes it difficult to understand why some knowledge enters the core and most does not. Thus it is on practical sociological grounds that I select my realist perspective.

My position on the epistemological issue has been strongly influenced by that of Elkana (1978), who suggests a way to combine the two philosophical approaches of realism and relativism. He claims that there is no way to determine empirically which of two broad conceptual visions or orientations is closer to the "truth." In this sense he is a historical relativist. But within a particular orientation some explanations are more supported by the empirical data than are others. Thus, for example, once one accepts the assumptions that are involved in doing sociological survey research, some interpretations of survey data are more correct than others. If, in a study of social mobility, class of origin is not associated with adult income, then the hypothesis that class of origin is the prime determinant of where one ends up in the social hierarchy is not credible. But this conclusion does not mean that the assumptions which must be made to do survey research are necessarily correct, or that they are more correct than the contradictory assumptions which the ethnomethodologists would make in doing their work. Within the assumptions made by ethno-methodologists, some work will be more correct than others. In this sense Elkana is a philosophical realist.

My position may be clarified by discussing a concrete example: the attempt by medical scientists to develop a vaccine against the human immunodeficiency virus (HIV), which causes AIDS. If a vaccine is developed and all people who are given it do not become HIV posi-tive, then we know that the vaccine works. If the people who receive the vaccine do become positive for HIV, the vaccine does not work. If we consider the scientific goal the development of a vaccine for the virus that leads to AIDS, it becomes clear that a successful solution to that problem could not be socially constructed *independent* of the external world, that is, independent of the nature of the virus. The external world is not, however, the only determinant of what happens in this area of science. It is possible, for example, that a solution to the problem could be ignored or rejected by the scientific commu-nity.[15] The way in which social processes might lead to such an out-come will be analyzed in Chapters 7 and 8. It is also possible that there may be more than one solution to this problem. The Sabin and Salk vaccines were both successful solutions to the problem of developing a polio vaccine. Social processes may be crucial in de-termining which of two successful solutions will become the dominant one. A vaccine may also be partially successful, and the extent to which it is effective could be socially negotiable. But I would argue that no completely unsuccessful solution (one in which there are no differences between the experimental group and the control group), no matter who proposes it or what the interests are of the various actors, will come to be accepted by the scientific community. Nature poses some strict limits on what the content of a solution adopted by the scientific community can be. By leaving nature out, the social constructivists make it more difficult to understand the way in which the external world and social processes interact in the development of scientific knowledge. Bringing nature back in, not as the only or the invariably most important influence on the content of science, but as one important influence, enables the sociology of science to address what will be its most important problem: how social processes and evidence from the empirical world interact to produce specific knowledge outcomes.

If we pursue the example of AIDS research further, we see that from a longer time perspective the relativist position may have validity. Suppose that a successful vaccine is developed and AIDS is eliminated as a significant health hazard, just as polio has been virtually eliminated by the successful development and administration of vaccines. Despite the successful solution to the practical problem, the scientists' views of what causes the disease and the mechanism through which the vaccine works will be based upon theories. That the vaccine works does not necessarily mean that the theories upon which it is based will always be thought to be correct. A change in paradigm in biochemistry, virology, and the other sciences involved in AIDS research could lead to new theories which are incommensurable with those currently held.

A practical solution to this seeming contradiction may be found by viewing underdetermination as a variable. Scientists are constrained to develop theories which fit what is known about nature. For the sociologist studying science it makes almost no difference whether what is "known about nature" is "really" "true" or not. The point is that at any point in time there is a body of knowledge that scientists accept as being an accurate approximation of nature, for a new idea to be accepted it generally must mesh with this body of knowledge. Occasionally, in what Kuhn terms revolutions, ideas which clash with the existing body of knowledge are accepted. Sociologists and historians still know relatively little about the conditions which determine when revolutionary ideas will be accepted and when they will be ignored. But leaving revolutionary science aside and considering "normal" science (that is, virtually all of the science studied by social constructivists), it seems likely that constraint is imposed by the existing body of accepted knowledge in the same sense that positivists might say that constraint is imposed by nature. The accepted body of knowledge is the functional equivalent of nature.

I will argue that scientists evaluate discoveries according to their utility in generating new research puzzles and in solving existing ones. Although nature may not completely determine the content of a highly useful discovery, in some cases nature places strict limits on what is likely to be perceived as a useful solution. In fact it is possible that the existing body of knowledge, evidence from the empirical world, and the ways in which scientists are trained act together to put severe constraints on what is likely to be put forth as a solution to any scientific problem. These factors would work together to rule out the great majority of possible solutions; but they may not determine which among a small set of possible but different solutions is the "right" one. My position is that cognitive factors and social factors interact to determine the fate of a new contribution. Leaving out either set of factors will lead to a one-sided and unrealistic view of how science is actually done.

Social Influences on Evaluation

Even though the empirical world may have considerable influence in constraining the content of new scientific contributions, in the great majority of cases the content of these contributions will not be sufficient to determine how they will be evaluated. In addition to emphasizing the importance of cognitive variables in the evaluation process, I will analyze the extent to which evaluation is also influenced by the social characteristics of the scientists making the discoveries and by social processes such as the operation of intellectual authority. In Chapters 7 and 8 I examine how social processes and variables influence the evaluation of new scientific contributions. I report results of research on how the characteristics of both the scientists who make the contributions and the scientists who pass judgment on the contributions influence the evaluation process. This research has suggested that these characteristics as they have been studied by sociologists of science play a very limited role in how new scientific contributions are evaluated. Variables such as the gender, age, religion, or race of the scientists making discoveries have not been shown to have a significant influence on how new scientific contributions are evaluated. Similarly, variables such as the prestige rank of current academic department and indicators of past track record such as the number of citations to past work show only weak correlations with scores given to grant proposals, as will be discussed in Chapter 6.

In the past evidence such as low correlations between gender and scientific recognition led me to conclude that science was basically universalistic in the way in which it evaluated new scientific contributions and their authors (J. Cole and S. Cole, 1973). In this book, however, I will argue that this conclusion was based upon an overly narrow definition of particularism and an overly narrow mode of studying the question. Essentially the prior findings that the characteristics of the authors have little influence on the evaluation of their scientific contributions were valid, but these low correlations do not necessarily mean that science is universalistic. Although evaluations in the scientific community do not seem to be as strongly influenced by the characteristics of the authors as by the characteristics of the work itself, these evaluations are not necessarily "objective." At the research frontier there are generally no agreed-upon criteria or norms which enable scientists to reach consensus, and therefore all evaluations, even if based strictly upon the content of the work, are subjective. The evaluation is based upon the subjective intellectual tastes of the evaluator, tastes which may differ from those of the person whose work is being evaluated or from those of other evaluators. As I shall discuss in Chapter 8, it is impossible empirically to separate evaluation based upon subjective assessments of the *work* from the attitude of the evaluator toward the *person* whose work is being evaluated. Because such personal valences undoubtedly play a role in much evaluation, science must be considered to have a particularistic dimension. I also argue that, given the nature of frontier knowledge, subjective evaluation and this type of particularism are an inevitable, and even a necessary, part of the development of science.[16]

Finally, in considering how social factors influence the cognitive content of science we must make a distinction between *local knowledge outcomes* and *communal knowledge outcomes*. A local knowledge outcome is the work produced in a particular context by one or more scientist. A communal knowledge outcome is work which is accepted by the relevant scientific community as important and true. I accept the fact that local knowledge outcomes may be influenced by social processes and chance factors, but I do not believe that anyone has demonstrated that the content of communal knowledge outcomes is influenced by social variables and processes.

My studies of consensus have concentrated on how work is evaluated rather than on the social processes which influence scientists while they are producing their work.[17] The most important of these processes is the operation of intellectual authority. Each individual scientist does not evaluate for herself every new truth claim. Rather our views of what is true and important are influenced by the opinions of colleagues. Some colleagues have more influence than others. In establishing a working consensus, science depends upon a relatively small group of scientists whose opinions are respected and who serve formally or informally as intellectual authorities. Some of the evaluation done by authorities is formal and involves the distribution of scarce resources such as prizes, grants, and prestigious positions. Most of the evaluation done by authorities is informal. We do not know very much about the mechanisms through which authority works, and this question is one of the most important topics on the research agenda of the sociology of science.

Even if social processes such as the operation of intellectual authority do not determine the actual content of the discoveries which are accepted into the core, they do determine which discoveries among a group of contenders for the attention of the scientific community will be successful. The extent to which authorities base their evaluations on the attributes of the contributions they are evaluating or on the characteristics of their authors is unknown. Presumably authorities have some interest in promoting work which supports or fits in with their own and in supporting the work of students and friends. Authorities who promote work which others see as inferior, however, can lose their legitimacy.

To what extent is the content of science determined by social factors? We are only now beginning to do the kind of detailed empirical and theoretical investigations which will be necessary to answer this question. This book is intended to contribute to this effort by raising questions about the logic of the constructivist approach to the sociology of science. My tentative answer to the question is that chance and social processes have important influences on the foci of scientific attention, the rate of scientific advance, and the production of new contributions in the laboratory. But when we consider the actual content of *communally* accepted knowledge, or the core, these random and social processes do not fully explain it. The social variables interacting with cognitive variables do influence the foci of attention and rate of advance, but social variables cannot be used to explain why one model of DNA rather than another was accepted into the core. This position, then, although it rejects completely the rationalistic view of science, is a form of realism.[18]

The Plan of the Book

In Chapters 2 and 3 I critically analyze the work of sociologists of science who have attempted to study the cognitive content of science.

I conclude that there are no successful case studies which show social influences on science's specific content as opposed to its foci of attention or rate of advance. I have three main criticisms of the constructivists: that they fail to explain why some contributions are accepted into the core while others are ignored or rejected; that analytically they conflate social with cognitive influences, making their argument tautological; and they fail to show linkages between specific social processes and specific cognitive outcomes, relying instead on general arguments where specific demonstration is needed.

In Chapter 4 I begin the presentation of my quantitative research on consensus and evaluation in science. This chapter presents empirical evidence to support my conclusion that at the research frontier there are low levels of consensus. The analysis is based upon a study of the peer review system used by the National Science Foundation (NSF).

Chapter 5 examines the hypothesis that we should expect to find higher levels of consensus at the research frontier of the natural sciences than of the social sciences. The empirical evidence suggests that this hypothesis must be rejected and that level of consensus at the frontier is not influenced by either the type of phenomenon studied or the level of consensus on core knowledge. These findings support the view that the natural sciences are indeed underdetermined and do not develop in the rationalist fashion depicted in traditional textbooks in the philosophy of science.

In Chapter 6 I begin my analysis of the influence of social processes on the evaluation of scientific contributions. Data from the NSF peer review study are used to examine the extent to which the characteristics of grant applicants influence the ratings their grants receive. The correlation between applicant characteristics and grant ratings are much smaller than expected. Data are presented which suggest that even the moderate correlations observed might be a result of differences in the assessed "quality" of the proposals rather than a direct influence of applicant characteristics on the evaluators. In general the evidence suggests that applicant characteristics have very little influence on the evaluation of their work.

In Chapters 7 and 8 I address the problem of universalism in the evaluation system of science. The statistical evidence as previously analyzed suggested relatively low levels of particularism at the institutional level, but these data are in sharp contrast to qualitative observations which show scientific evaluation to be permeated with particularism. In order to resolve this paradox I critically examine both the way in which universalism has been conceptualized and the way in which it has been empirically investigated. The conclusion is reached that subjective and particularistic network ties based upon empirically overlapping cognitive and noncognitive interests play an important role in the evaluation of scientific contributions.

Although the sociology of science is capable of telling us a great deal about how social and cultural factors influence the rate of scientific advance, very little work has been done on this topic. Chapter 9 reviews the ways in which sociologists have studied the rate of advance and presents data from a study which suggests how research in this direction may be fruitfully conducted.

In Chapter 10 I summarize the book's conclusions and analyze the problems that would be encountered in attempting to develop an empirical test of some of the main hypotheses raised by constructivist sociologists of science. I conclude that some aspects of their program could be investigated using systematic quantitative studies. For other parts of their research program systematic empirical investigation is not feasible. Constructivist sociology of science has raised a whole series of interesting questions. I propose that in the future constructivist and nonconstructivist sociologists of science collaborate in attacking those questions which we can solve, and leave to others epistemological issues which cannot be resolved by sociological methods.

· · 2 · ·

Constructivist Problems in Accounting for Consensus

Prior to the 1970s sociologists of science paid very little attention to the cognitive content of science. Today the attempt to show social influences on the content of the natural sciences is the major focus of the discipline. Do we now have any evidence to reject Mannheim's view that the content of the natural sciences is not on the whole susceptible to social explanations?

By the content of science I mean the actual substance of the scientific ideas that are developed in the laboratory and then evaluated by the scientific community. As pointed out in Chapter 1, I distinguish between the content of scientific ideas on the one hand and the rate of scientific advance and the foci of scientific attention on the other. It is clear that the latter are influenced by social conditions, but what about the former? Do we have any evidence that as a result of social factors the cognitive content of science is different from what it otherwise might have been? Although much recent research in the sociology of science has provided useful descriptions of how science is actually done, it has failed to show how cognitive dependent variables are influenced by social independent variables.

In this and the following chapter I present a critique of the research program of the social constructivists. When this approach to the sociology of science began to emerge in the 1970s it was practiced almost exclusively by a small group of Europeans who saw themselves as Davids challenging the Mertonian Goliath. Opposition to Mertonian sociology served to hold the constructivists together. But now that the constructivist approach has become the dominant one, many differences and schisms have developed within the school.[1] The various branches of constructivism are discussed informatively but in quite different ways by Zuckerman (1988) and Ashmore (1989).

One group, that which I find the most interesting, has based its analysis on participant observation of scientists' behavior in the laboratory (Knorr-Cetina, 1981; Latour and Woolgar, [1979] 1986; Lynch, 1985). Members of this group try to show how the products of laboratory work, local knowledge outcomes, are heavily influenced by specific aspects of the context in which they are produced. They point out that the work they observe is significantly underdetermined by data and external "reality" (if such exists) and heavily influenced by chance and social negotiation. As we shall see later in this chapter, laboratory studies have been criticized both by other constructivists and by nonconstructivists for not adequately dealing with the communal evaluation of local knowledge outcomes.

Other constructivists use traditional historical methods and in-depth interviews to conduct case studies of the development of particular research areas, usually with an emphasis on conflict resolution (Collins and Pinch, 1982; Pickering, 1984; Pinch, 1986). They focus on how the choice between competing solutions to a scientific problem is influenced by social factors. One of the primary explanatory schemes used is the "interests" model. It is argued that which solution wins out is heavily influenced by both the cognitive and the social interests of various members of the scientific community. This theory will also be examined in this and the following chapter.

Some constructivists have become interested in "discourse analysis," which is based on analysis of both the conversation and the writing of scientists (Gilbert and Mulkay, 1984; Potter, 1984; Yearley, 1981). This work points out that the accounts given by different scientists of the same scientific ideas and experiences vary substantially. It is then claimed that until we understand the sources of this variability, we cannot use the statements of scientists about their own work as data to understand what is actually going on. This approach is then critical of the work of other constructivists who do make use of such accounts. Most recently constructivists with an ethnomethodological orientation, such as Woolgar and Ashmore, have focused on the problem of reflexivity, applying constructivist principles to an analysis of the work of scholars conducting research in the sociology of science (Ashmore, 1989; Mulkay, 1985; Woolgar, 1988). Reflexivity has in the past been a problem for constructivists who do any type of empirical research. If, in fact, scientific decisions are not influenced in important ways by empirical evidence and logical argument, why do proponents of this view bother conducting empirical studies? Ashmore (1989), in particular, tries to show how reflexivity, rather than being a problem, offers constructivist sociologists new and interesting opportunities.

All of these approaches are micro-level in the sense that they study detailed behavior in the laboratory or detailed analysis of specific texts and conversations. There are also members of the constructivist school, practitioners of the the "strong program" (to be discussed in Chapter 3), who take a more macro-level approach and investigate the influence of the broader social context or interests of scientists on their cognitive work (Barnes and MacKenzie, 1979; Bloor, 1976, 1982; MacKenzie, 1981; Shapin, 1982; Shapin and Schaffer, 1985). Thus Bloor (1982) argues that Boyle's class interests influenced his theory of gases. And, finally, some macro-oriented constructivists have attempted to use traditional quantitative sociological methods to collect empirical evidence on the extent to which science is socially influenced (Stewart, 1990).

The Core of Constructivism

Three broad assumptions characterize virtually all constructivist writing. First, all constructivists dispute the traditional view of science as a uniquely rational activity. Second, almost all constructivists have adopted a relativist epistemological position which emphasizes the underdetermination of solutions to scientific problems and deemphasizes or altogether denies the importance of the empirical world as a constraint on the development of scientific knowledge. Third, all constructivists argue that the actual cognitive content of the natural sciences can only be understood as an outcome of social processes and as influenced by social variables. As I pointed out in Chapter 1, I am in agreement with constructivists on the first assumption. On the second assumption, although I agree that science is underdetermined, I reject relativism and place a much stronger emphasis on constraints posed by the empirical world. On the third assumption I see the foci of attention, the rate of advance, and the everyday doing of science as influenced by social processes; but thus far there has been no research which presents a convincing case that the actual content of communal knowledge outcomes is determined by these social processes.

It should be clear from the programmatic statements made by the constructivists that they are not primarily interested in either the foci of scientific attention or the rate of scientific advance. By the cognitive content of science they mean the actual substance of scientific ideas. Consider the reply of the constructivist Harry Collins to the critique of this program made by Gieryn (1982), who argued that Merton actually anticipated much of the work of the constructivists.[2] In reply, Collins argues that Merton never claimed that social variables could explain the actual content of scientific ideas:

> Even Merton's programmatic statements as quoted by Gieryn do not begin to be mistakable for an anticipation of the modern programme unless Merton intended to say not just that the conditions for the existence of scientific practice are social, and not just that the direction and focus of attention of science is in part socially determined, but also that the very findings of science are in part socially determined. The relativist programme tries to do analyses that imply that *in one set of social circumstances "correct scientific method" applied to a problem would precipitate result p whereas in another set of social circumstances "correct scientific method" applied to the same problem would precipitate result q, perhaps, q implies not-p.* (1982, p. 302, italics added)

I agree with Collins that Merton never thought that sociology would be able to conduct the kind of analyses to which Collins refers. But have the social constructivists actually given us such analyses? It seems clear from this statement by Collins that he and his colleagues want to demonstrate the social determination of very specific aspects of the content of science; they mean the social determination not just of general orientations or of general cognitive styles but of the detailed substance of what is in scientific papers.

There now exists a large body of work by sociologists following this perspective. The three works which I find to be the best exemplars of constructivism are Bruno Latour and Steve Woolgar ([1979] 1986), *Laboratory Life: The Construction of Scientific Facts;* Bruno Latour (1987), *Science in Action;* and Karin D. Knorr-Cetina (1981), *The Manufacture of Knowledge: An Essay on the Constructivist and Contextual Nature of Science.*

The Latour and Woolgar monograph is based upon two years of field work conducted by Latour at the Salk Institute in San Diego, California. Latour took the approach of an anthropologist studying an unknown tribe. Claiming to have abandoned all assumptions about science, his goal was to present an account of how science was conducted based on the observed behavior of the scientists rather than on any preconceptions he might have about science. The book contains a detailed historical analysis of the "discovery" of TRF (or TRH).[3] The structure of this chemical peptide, Thyrotropin Releasing Factor (or Hormone), a substance released by the thalmus, was "discovered" almost simultaneously by R. Guillemin (Latour's host at the Salk Institute) and A. V. Schally, two scientists who later shared the Nobel Prize for their work.[4] This case study is based upon traditional techniques, including an analysis of the published literature, analysis of citations to this literature, and in-depth interviews with key participants. Latour's book *Science in Action* (1987) is not based upon laboratory observations but contains a detailed analysis of a series of cases, including the development of a computer at Data General, the introduction of the Diesel engine, and the discovery or social construction of Growth Releasing Factor (GRH) by Guillemin. Knorr-Cetina's book is based upon field work she conducted in a protein chemistry laboratory at the University of California at Berkeley. Like Latour she observed scientists at work, conducted in-depth interviews with them, and analyzed their publications.

All three monographs have similar theses. Starting with the assumption that science is underdetermined, they argue that in the laboratory the role of the empirical world or nature is so far removed that scientists have a great deal of freedom to construct scientific knowledge in different ways. Even if nature exists outside the laboratory, for all practical purposes work in the laboratory cannot be said to be influenced by nature. In fact, rather than nature being of some influence in determining the content of scientific theories, it is the behavior of scientists in doing science which defines what nature is. The causal order between nature and the content of science has been reversed.

Because the ideas developed by scientists in the laboratory cannot be understood as being constrained by nature, it is possible to show that social processes influence the formation of these ideas: "The thesis under consideration is that the products of science are contextually specific constructions which bear the mark of the situational contingency and interest structure of the process by which they are generated, and which cannot be adequately understood without an analysis of their construction. This means that what happens in the process of construction is *not* irrelevant to the products we obtain" (Knorr-Cetina, 1981, p. 5).

What exactly is meant by social construction? If there are any existing laws of nature, when scientists are working at the frontier on unsolved problems (as they always are), the content of the law or fact to be discovered will be completely unknown. Traditional views of science that empirical evidence will enable the scientist to decide if an idea is right or wrong are an oversimplification at best and completely incorrect at worst. In trying to make sense out of a confusing mass of almost endless possibilities, scientists make arbitrary decisions, decisions which could easily have been made in some other direction. As scientists try to develop meaningful frameworks to interpret their data they are influenced by social interaction, directly with colleagues in the laboratory and indirectly through the literature. The decisions which scientists make are a result of social negotiations which take place both outside and inside the laboratory. The constructivists claim that we can only understand the work which scientists produce by studying the social processes that influence its creation.

As I interpret their thesis they view the specific content of scientific work (or "productions," as the constructivists call them) as the dependent variable.[5] Thus in the constructivist research program the content of science is not treated as a "black box." The independent variables are the social processes that influence the work of scientists in the laboratory. Presumably some social processes lead to one cognitive outcome, whereas other social processes would lead the same scientists working on the same problem to produce another cognitive outcome. That *social* reality is constructed has long been the position of phenomenologically influenced sociologists and social psychologists (Berger and Luckmann, 1963); it is one of the premises of the research program of the ethnomethodologists in contemporary sociology. What is unusual about the work done by the constructivists in the sociology of science is their application of this perspective to the natural sciences. They want to convince us that *natural* reality, as well, is socially constructed.[6]

In this chapter I focus on the problems which face adherents of the constructivist approach in trying to explain why one idea wins out over others in the competition to become part of the core. The constructivists are not, in my view, able to explain adequately why a few scientific ideas are taken up by the community and become universally accepted, whereas others move quickly to obscurity. The concentration of many constructivists on micro-level analyses leads them to underemphasize the significance of what happens to scientific findings when they leave the laboratory.[7] This micro-level emphasis, coupled with their relativistic perspective, causes them to underestimate the extent to which the behavior of scientists in the laboratory is constrained both by the natural world and by the cognitive norms that currently exist in the relevant scientific community.[8]

Moving from the Frontier to the Core

One of the basic challenges faced by the sociology of science is to understand why some science produced at the research frontier moves into the core, some work is seriously considered as a candidate for the core but is ultimately rejected, and most work is ignored. How does a local knowledge outcome become an accepted communal knowledge outcome? Understanding the ways in which new scientific contributions are evaluated is the key to understanding the social influences on community-based scientific knowledge. Traditional positivists could answer this question easily. That science which was supported by the evidence of nature would be added to the core. The constructivists underestimate or completely deny the significance of the empirical world and depend instead on the importance of social strategies employed by scientists to convince others that their work should be accepted. I argue that it is impossible to explain which new contributions are accepted by the community without considering characteristics of the contributions themselves. Among the most important of these characteristics are the extent to which the new contribution is supported by empirical evidence and fits in with what is currently held to be true and important by the community.

Let us consider first how the constructivists *describe* (as opposed to *explain*) the change which occurs in science when it moves from the frontier to the core. Latour and Woolgar ([1979] 1986) point out that from looking at textbook science one gets the impression that science is made up of facts which have been discovered. Once scientific ideas are accepted by the scientific community they take on a "facticity" which make them seem to be part of nature, something which was inevitable. At any given time there are a series of assumptions or taken-for-granted facts (what I call the core) accepted by scientists as true without question. This core knowledge differs significantly from the status of the work the scientists are currently doing, which is questioned and not taken for granted. But at some time in the past those assumptions or facts which are now accepted without question must have occupied a status similar to the work which the scientists are currently doing: "It was obvious to our observer, however, that everything taken as self-evident in the laboratory was likely to have been the subject of some dispute in earlier papers. In the intervening period a gradual shift had occurred whereby an argument had been transformed from an issue of hotly contested discussion into a well-known, unremarkable and uncontentious fact" (p. 76).

In *Science in Action* Latour (1987) develops further his argument about the Gestalt-like switch which takes place as scientific statements move up the "ladder of facticity." At the lower levels of the ladder the statements are seen by scientists as the products of particular investigators and thus subject to questioning and challenge. But when the same statement reaches the top of the ladder the scientists see the statement as being determined by "nature," not by the work of men and women. At this point scientists use what Gilbert and Mulkay (1984) have called the "empiricist repertoire" to describe the fact as made. "As long as controversies are rife, Nature is never used as the final arbiter since no one knows what she is and says. But *once the controversy is settled,* Nature is the ultimate referee" (Latour, 1987, p. 97).

The constructivists' description of the change which knowledge undergoes as it moves up the "ladder of facticity" corresponds closely to my distinction between the frontier and the core. But the constructivist position does not explain why some discoveries are able to make this transition while most are not. Sociologists of science are in agreement that in order for a scientific contribution to achieve the status of fact it must be accepted by its audience, the scientific community. Latour and Woolgar ([1979] 1986) recognize the importance of the reaction of members of the scientific community to products produced in the laboratory. Positive reaction can increase the facticity of a finding and negative reaction can decrease it. They give an example from a referee's report on an article submitted for publication by one of the lab scientists: " 'The conclusion that the effect of Pheno . . . [to] release PRL *in vivo* is mediated through the hypothalmus *is premature.*' Three references were then given, which further pulled the rug from under the author's conclusion" (p. 84).

In *Science in Action* Latour (1987) emphasizes that the fate of a scientific statement is wholly in the hands of the community. No matter what the "objective" content of a scientific article, if it is ignored it cannot become a fact and will have no meaning for the growth of knowledge: "We have to remember our first principle: the fate of a statement depends on others' behaviour. You may have written the definitive paper proving that the earth is hollow and that the moon is made of green cheese but this paper will not become definitive if others do not take it up and use it as a matter of fact later on. You need *them* to make *your* paper a decisive one . . . The construction of facts, like a game of rugby, is thus a collective process" (p. 104).

Thus the results which are socially constructed in the laboratory are subject to scrutiny by members of the scientific community outside the laboratory. These outsiders are not influenced by the same micro-level context, nor do they necessarily have the same interests. Their interests may be in opposition to those of the scientists who published the paper they are reading. They may be competitors in a race to find a solution to a problem which will be generally accepted by the scientific community. If outsiders do not accept the validity of the ideas produced in the laboratory, these ideas are unlikely to become part of the communal body of knowledge. But because the social constructivists concentrate on micro-level contexts, they frequently do not have any data on the processes at work outside the laboratory which determine the fate of the statements produced within.

Studies based upon laboratory observation can give us an example of how the end product of a particular laboratory is socially constructed; but these studies are not effective in enabling us to understand the processes and variables which influence the reaction of the community of scientists to locally created products.[9] This point is now accepted by some constructivists. Consider this statement of Pinch (1986):

> The difficulty with this location [the laboratory] is that it is rather restrictive when it comes to the study of the process of consensus formation. This is because the group amongst which consensus forms is not situated in any single laboratory. They are typically scattered amongst a number of laboratories. This means that a rather one-sided picture of events will emerge from within a single laboratory. This contrasts with the controversy location where the focus on a wider group of scientists enables processes of consensus to be studied in a more convincing way. (p. 30)

Realizing that one must leave the laboratory to examine the social processes influencing consensus formation, some constructivists such as Pinch (1986) and Pickering (1984) have conducted qualitative case studies of how controversy was handled among a group of specialists. These studies have suggested the importance of both cognitive and social factors in resolving disputes. Pinch (1986, p. 175), for example, concludes that the consensus reached that an experiment was correct was "a social achievement." Although data were available which would have supported two different interpretations, the credibility of the experimenter was so high that others were willing to accept his interpretation and reject that of the critics, whose credibility was lower. Thus Pinch is giving an example of how the characteristics of the author of a scientific paper influence its reception. This type of research aims not at showing how social variables influenced a local knowledge outcome to take a particular form rather than another but at why others are willing to accept a local knowledge outcome.

Given the fact that interests and social contexts will differ for the readers of a paper and its authors, what factors determine the reaction of the readers to the paper? If the local knowledge outcome is defined by the community as trivial and is ignored, all the social processes and contextual conditions that influenced the production of the paper will have been irrelevant for the growth of communal knowledge in this area. If a paper is ignored it can have no impact on communal knowledge. An ignored paper has no more impact than an unpublished one. But suppose that the paper is not ignored; suppose that it turns out to be of at least some significance for the growth of knowledge. Why, we must ask, will one paper produced by one group of scientists, influenced by a particular set of social circumstances, be accepted as more credible or more important than another paper published by another group of scientists influenced by another set of social circumstances?

Rejection of Replication

Positivists try to deal with this problem by pointing to the role of replication in science. If a discovery is important it will be replicated by other scientists working in other contexts. If it is successfully replicated it will move up the "ladder of facticity." As pointed out in Chapter 1, the constructivists deny the significance of replication, citing the work of Collins (1974, 1975). Although Collins' analysis is based upon a case study of only one research area in physics, the constructivists have concluded that replication is very rarely, if ever, done in science and that even if one wanted to replicate an experiment it would be very difficult because replication depends upon tacit knowledge (knowledge which is not conveyed in journal articles) rather than following a set of rules of procedure.

Although exact replication may not be common in science, dependence upon *confirmation* of results by others is. For example, the Gargamelle collaboration at the international research center CERN, in Geneva, discovered neutral currents in an experiment in 1973. Not everyone in the world of high-energy physics was convinced by the experiment. The results could have been an artifact. Indeed, researchers in the E1A experiment at the Fermi National Accelerator Laboratory in Batavia, Illinois, who had also been searching for neutral currents, at first failed to identify them. The E1A was not a replication of the Gargamelle. Rather it was an attempt to identify the same phenomenon using a different experimental procedure. At first the data from E1A convinced the physicist David Cline of the University of Wisconsin that his long-standing belief that neutral currents did not exist was correct. But additional analysis of the data showed Cline and his associates that their experiment had also found evidence for neutral currents: "'Three pieces of evidence now in hand point to the distinct possibility that a [muon]less signal of order 10% is showing up in the data. At present I don't see how to make these effects go away'" (quoted in Galison, 1987, p. 235). When the Fermi lab team confirmed that they too had found evidence for the neutral currents, the rest of the high-energy physics community began to find the evidence more convincing. If Cline and his E1A collaborators had continued to argue against the existence of the neutral currents and displayed data which backed up their argument, the claim of the Gargamelle group would have been further from being accepted as fact.[10] The importance of experimental confirmation in establishing consensus is ignored or underemphasized by the constructivist sociologists.

External Factors

In trying to address the problem of why some work is accepted and other work is ignored or rejected by the community, Knorr-Cetina (1981) argues that there is no distinction between how social processes work in the laboratory and how they work in the scientific community in the evaluation process. She claims that scientists in the laboratory are influenced by their knowledge of the criteria utilized in the community:

> Scientists constantly relate their decisions and selections to the expected response of specific members of this community of "validators," or to the dictates of the journal in which they wish to publish. Decisions are based on what is "hot" and what is "out," on what one "can" or "cannot" do, on whom they will come up against and with whom they will have to associate by making a specific point. In short, the discoveries of the laboratory are made, as part and parcel of their substance, *with a view toward* potential criticism or acceptance (as well as with respect to potential allies and enemies!). (p. 7)

If this description of the behavior of scientists is accurate, we may conclude that there are rules of procedure *external* to the laboratory that the scientists are more or less aware of and that influence the decisions they make in the laboratory. Thus if what the scientist does in the laboratory is not constrained by the empirical world, it is at least constrained by the scientist's knowledge of the criteria by which her work will be evaluated. Although constructivists generally want to deemphasize external factors, they are forced to import them when considering the question of how the laboratory productions are evaluated by the community.

Knorr-Cetina makes a further point in suggesting an analogy between the selection of ideas over time and selection in biological evolution. "Like adaptation, acceptance can be seen as the result of contextual pressures which come to bear on the scientists' selections in the environmental niches provided by the laboratories" (1981, p. 9). But in the case of biological selection in evolution, the mechanism of survival is adaptation to *objective* conditions which determine whether a particular individual of a species will survive and reproduce. These conditions are external to the individuals. Knorr-Cetina does not specify any mechanism which could drive selection in science.[11]

Consensus Based on "Strategies"

If replication and confirmation are not the means through which communal consensus develops, the problem of explaining how the community does reach consensus remains. Because the constructivists reject the answer of replication, they need an alternative solution to this crucial problem. Latour and Woolgar ([1979] 1986) and Latour (1987) emphasize the *strategies* that scientists use in order to convince others that their statements should be accepted as facts. The strategies consist of the use of rhetoric, in fictionalizing what went on in the laboratory into a coherent story, and the enlisting of allies. To Latour (1987) the writing of scientific papers is only on the surface an attempt to convey to the community results achieved in the laboratory. The scientific paper is in fact a rhetorical device with the *real* aim of gathering supporters for the author's views and challenging the views of competitors.

Consider Latour's analysis of the function of citations in scientific literature. The citation of other papers is used as part of the rhetoric and as an attempt to line up the literature on one's side and, in some cases, against one's scientific opponents: "A given paper may be cited by others for completely different reasons in a manner far from its own interests. It may be cited without being read, that is perfunctorily; or to support a claim which is exactly the opposite of what its author intended; or for technical details so minute that they escaped their author's attention; or because of intentions attributed to the authors but not explicitly stated in the text; or for many other reasons" (p. 40).[12]

Latour *describes* the strategies used by scientists in attempting to convince others to take up their ideas and use them in their own work, but he has very little to say about what determines success. From his emphasis on strategies and rhetoric one might conclude that he believes the determining factor to be the skill of the scientist in employing these devices, *independent* of the content of their papers. But without considering the content of what scientists say in their papers, we cannot explain why some scientists succeed and others fail with their rhetorical strategies. Consider again the famous paper by Watson and Crick putting forth the double helix model of DNA. The constructivists ask us to believe that if a paper containing the same model had been written up in a different way it would not have been accepted.

If we examine the writings of the constructivists carefully we will see that even they are aware that there may be more than rhetorical style and strategies which determine the community's reaction to scientific work:

Participants were more convinced that an inscription unambiguously related to a substance "out there," if a similar inscription could also be found. In the same way, an important factor in the acceptance of a statement was the recognition by others of another statement which was similar. The combination of two or more apparently similar statements concretised the existence of some external object or objective condition of which the statements were to be taken to be indicators. Sources of "subjectivity" thus disappeared in the face of more than one statement, and the initial statement could be taken at face value and without qualification. (Latour and Woolgar, [1979] 1986, pp. 83–84)

Although the constructivists have attempted to show that replication is generally not done in science, in the preceding quotation we see the recognition by Latour and Woolgar of the importance of confirmation. If similar or confirming results are obtained in two different experiments in the same laboratory or from experiments in different laboratories, these results become more believable and are more likely to move up the ladder of facticity. What variables determine the chances that a particular scientific finding will be confirmed or replicated? Successful replication or confirmation might simply be a matter of chance *or* that some findings are more likely to correspond with the natural world.

Latour (1987) gives us additional insight into what the characteristics of convincing articles might be. He describes scientific procedures such as "the cartilage tibia assay, Weber's first antenna and the liquid microbe culture," which generate results which are subject to multiple and conflicting interpretations. If a scientist utilizing these procedures "wanted to be believed he would have an endless task. Every time he opened his mouth, any number of his dear colleagues would start shaking their heads and suggesting many alternatives just as plausible as the first. To do so, they would only need a bit of imagination" (p. 82). Scientists can gain the credibility they seek for their statements by developing a much finer measuring device, one which will generate results that are more difficult to dispute:

But the way out of the controversy that interests us here is to build *another* antenna, one, for instance, that is a thousand million times more sensitive than Weber's so that this part of the detection at least is not disputed. The aim of this new antenna is to confront the skeptic with an incontrovertible black box *earlier* in the process. After this, skeptics may still discuss the amount of gravitation, and what it does to the relativity theory or to astrophysics, but they will not argue that there are peaks that cannot be explained by terrestrial interferences. With the first antenna alone, Weber might be the freak and the dissenters the sensible professionals. With the new antenna, those who deny the presence of the peaks are the isolated skeptics and it is Weber who is the sensible professional.[13] (Latour, 1987, p. 82)

When the new antenna or other similar "black box" is constructed, the audience for the paper, regardless of the contexts in which they work or the interests that they have, will have difficulty disputing its results. Why do people who could so easily reject the results from the weaker antenna *have* to accept the results from the stronger? Some instruments and some experiments yield results which more easily exclude alternative interpretations. Whether these results do so because they conform more closely to nature or because they better meet the standards that the discipline has for making evaluations really does not matter. What does matter is that for each research area there will be some types of results which will be more likely to convince than others. The cognitive productions have attributes which influence their reception as much as or more than does the rhetorical style in which they are presented or the social context of the producers or the evaluators.

Utility and the "Diffusionist" Approach

The attributes of contributions which influence their reception will differ from research area to area and over time. Let us call these attributes the perceived *utility* of a scientific contribution.[14] I believe that because of the openness of science and its highly decentralized structure the utility of a scientific idea will have a significant influence on its fate. Utility has two general components. First, an idea is useful if other scientists can build upon it or use it in their own work. This type of idea is in Kuhn's terms "puzzle generating." Second, an idea is useful if it works, that is, if it generates results which are expected or which can fit in with other scientific ideas. We may call such ideas "puzzle solving." An idea can have high utility if scientists perceive it to have the *potential* to generate solutions to many scientific problems. Some scientific contributions, such as the chemical sequence of TRF, were both puzzle generating and puzzle solving.

Latour (1987) decisively rejects the approach that I take, calling it "diffusionist":

First, it seems that as people so easily agree to transmit the object [such as the structure of DNA after discovery by Watson and Crick], it is the object itself that forces them to assent. It then seems that the behaviour of people is *caused* by the diffusion of facts and machines. It is forgotten that the obedient behaviour of people is what turns the claims into facts and machines; the careful strategies that give the object the contours that will provide assent are also forgotten. Cutting through the many machiavellian strategies of this chapter, the model of diffusion invents a technical determinism, paralleled by a scientific determinism. Diesel's engine leaps with its own strength at the consumer's throat, irresistibly forcing itself into trucks and submarines, and as to the Curies' polonium, it freely pollinates the open minds of the academic world. Facts now have a *vis inertia* of their own. They seem to move even without people. More fantastic, it seems they would have existed even without people at all. (p. 133)

I plead guilty to being a diffusionist: not the extreme diffusionist which Latour caricatures, but a diffusionist nonetheless. Latour does not explain why at some times people behave "obediently" and at others they behave "disobediently." Without using characteristics of the "object" it seems impossible to explain very much of the variance on this crucial question. A related question is why some scientific contributions are almost immediately taken up by the scientific community as facts, and others have to fight a long and hard battle to achieve that status. Contributions that are immediately accepted pose a particular problem to the argument that acceptance by the scientific community cannot be in any way explained by the cognitive characteristics of a contribution. If we consider an invention such as the Diesel engine, which did not work at first and took many years to be developed to a state which was accepted, it is easy to agree with Latour that there was nothing about the content of the technological discovery which forced people to accept it. But a case such as Watson and Crick's presents a more difficult challenge.

When a new contribution is almost immediately given the status of fact and added to the core it is hard to avoid the conclusion that there probably was something about its cognitive content, independent of rhetoric or other social means of persuasion, which played some role in convincing the members of the audience to accept it. I am arguing not that social processes have no influence on the evaluation of new contributions, but that cognitive attributes play at least some role in the evaluation. Even in the case of the DNA model, which took a very short time to enter the core, the truth of the model was not completely independent of the social context. How did Watson and Crick know that they had discovered the structure of DNA? They thought they had it, but several times before they had jubilantly proclaimed that the structure had been discovered only to have their enthusiasm dashed by the criticism and skepticism of some of their colleagues, particularly Maurice Wilkins and Rosalind Franklin. Thus when they did come up with the model that was to be accepted as the structure of DNA, they did not rush to publication before presenting it for the critical inspection of a small group of their peers. Only when the model passed muster with this group did they move to print.

Watson (1968) describes the reaction of several of his competitors upon seeing the final DNA model:

> Rosy's [Rosalind Franklin] instant acceptance of our model at first amazed me. I had feared that her sharp, stubborn mind, caught in her self-made antihelical trap, might dig up irrelevant results that would foster uncertainty about the correctness of the double helix. Nonetheless, like almost everyone else, she saw the appeal of the base pairs and accepted the fact that the structure was too pretty not to be true. Moreover, even before she learned of our proposal, the X-ray evidence had been forcing her more than she cared to admit toward a helical structure. The positioning of the backbone on the outside of the molecule was demanded by her evidence and, given the necessity to hydrogen-bond the bases together, the uniqueness of the A-T and G-C pairs was a fact she saw no reason to argue about. (p. 210)[15]

And:

> Pauling's reaction was one of genuine thrill, as was Delbrück's. In almost any other situation Pauling would have fought for the good points of his idea. The overwhelming biological merits of a self-complementary DNA molecule made him effectively concede the race. He did want, however, to see the evidence from King's before he considered the matter a closed book. (pp. 217–218)

There are some interesting discrepancies in the accounts given of this discovery by Watson (1968) and Crick (1988). Watson claims that as soon as they decided to publish the paper he was sure that the Nobel Prize would be theirs. Crick (1988) claims to be surprised by Watson's talk of the Nobel Prize and gives an impression of less certainty over the significance of the discovery. "Although we were excited when we discovered the double helix, neither we nor anybody else thought of it as a wild success. Indeed Jim worried that it might be all wrong and that we'd made fools of ourselves" (p. 87). Crick points out that there were some who were not enthusiastic over the double helix. Barry Commoner "insisted, with some force, that physicists oversimplified biology" (p. 77) and Erwin Chargaff found the first *Nature* paper to be "interesting" but did not agree with the second paper, in which Watson and Crick analyzed the genetic implications of their discovery.

Crick does say, however, that "we realized almost immediately that we had stumbled onto something important . . . I do recall going home and telling Odile that we seemed to have made a big discovery" (p. 77). Sir Lawrence Bragg, the head of the Cavendish Laboratory where Watson and Crick were working, was "immediately enthusiastic" (p. 77) after seeing the model. Crick also reports that just a few months after the publication of the initial paper the DNA model had been recognized as very important by the field's leading authority, Max Delbrück: "There is no doubt, however, that it had a considerable and immediate impact on an influential group of active scientists. Mainly due to Max Delbrück, copies of the initial three papers were distributed to all those attending the 1953 Cold Spring Harbor Symposium, and Watson's talk on DNA was added to the program" (p. 76).

Whether or not the discoverers of the double helix knew immediately that they would win the Nobel Prize, it was clear to them and all concerned that they had made an "important" discovery. The double helix model was made much more credible by the X-ray evidence which was published simultaneously by Maurice Wilkins and Rosalind Franklin. Whereas Watson had been hesitant to include a discussion of the genetic implications of the model in the original paper, when he saw the evidence from the X-rays he agreed to write another paper with Crick. Rather than saying that the Watson and Crick paper alone compelled acceptance, it would probably be more accurate to say that the simultaneous publication of the Watson and Crick paper with the two papers reporting the X-ray evidence compelled most authorities in the field to accept the validity of the double helix model—even those who were competitors of Watson and Crick and had an interest in their model being wrong.

I am not saying that the discovery of the model of DNA can be explained by its correspondence with nature. Anyone reading Watson's (1968) famous account of the discovery of the double helix will learn that Nature never spoke to the scientists. The discovery was made in part as a result of a series of chance factors which happened to put the right people together at the right time. But the primary question here is not how Watson and Crick *made* the discovery, but why it was so quickly *accepted* by others as a fact. At the time of the publication of the brief paper in *Nature* Watson and Crick were young scientists without any significant reputation.[16] That the paper was well written is hardly likely to account for its rapid acceptance. Otherwise all young scientists would have to do to become eminent would be to hire skilled editors to rewrite their papers.

It is important to remember, moreover, that the structure of DNA had been widely viewed at the time as one of the most significant problems in biochemistry. It was believed that whoever solved this problem would win the Nobel Prize. Watson (1968) claims that this was part of his motivation in pursuing the problem, and it is not surprising that there were many famous scientists, including Wilkins, Chargaff, and Linus Pauling, who were also at work on it. When the Watson and Crick paper was published, the other scientists working on the DNA problem were motivated *not* to accept it as a fact but to find something wrong with it or to find it unconvincing. Their career interests motivated them to reject rather than accept the Watson and Crick model. When they could not find anything wrong with it, and indeed it seemed to all to be true, they had to accept it and it became a fact. The acceptance as opposed to the discovery was not a result of rhetoric, contextual conditions, or special machinery available to Watson and Crick. The conclusion that there was something about the content of their model which influenced its rapid acceptance by others does not seem to be farfetched.

It is certainly true that after scientists such as Pauling and Delbrück had accepted and incorporated the Watson and Crick model in their work it became much more difficult for anyone to challenge the model. As Kuhn and others have pointed out, once a theory or paradigm has been accepted by the scientific community that paradigm resists change. But this does not explain why the most eminent scientists accepted the model in the first place.

There is one study which systematically addresses the question of why some contributions are more quickly accepted. John Stewart (1990) examines the emergence of the plate tectonics paradigm in geophysics. Stewart, who was trained as a traditional sociologist of science, has only one foot in the constructivist camp. He is sympathetic to the constructivist approach and adopts many of their ideas, although he is not a radical relativist. The first part of his book is based upon interviews and examination of documents. In Chapter 7 of his study he presents the results of quantitative analyses aimed at understanding the factors influencing the evaluation of a sample of geophysics articles. Stewart uses as the measure of his dependent variable the number of citations each article received in the five years after its publication. He then tries to predict this by using two sets of independent variables, characteristics of the article (which he refers to as intellectual resources) and characteristics of the author(s) (which he refers to as social resources). He finds that the characteristics of the articles explain approximately twice as much variance on the reception of the articles as do characteristics of the authors, and he concludes that intellectual resources were the most important influence on evaluation and consensus formation for geophysics during the period of time that he studied.

The measures which Stewart used to characterize the articles are important. The variable with the greatest influence on citations was the length of the article. Also influential was the extent to which the article cited work from either of the two most significant research approaches (as identified by a co-citation analysis). Thus those articles which cited key plate tectonics articles were themselves more likely to be cited than those which did not cite the cornerstones of the plate tectonics paradigm. Other variables which had a significant influence were the number of references to articles published in the last three years, the length of period between acceptance of the article and its publication (negative effect), and the extent to which the article used empirical data. Although these variables are strongly correlated with the reception of the new articles, they tell us very little about their content. Would one want to argue that if the real content of the article or its "quality" were held constant, that simply increasing its length or the number of recent references would lead to a more favorable evaluation? There is a strong possibility that the effect of most of these variables would be eliminated if we had better measures of the real content of the articles. Consider, for example, the influence of citing key paradigm publications on the reception of the new article. We would not expect an article citing the same set of references as the famous Watson and Crick article but not introducing a credible model for DNA to have received the same reception as did the Watson and Crick article. And would we expect a paper which reported a result such as the Watson and Crick model but failed to cite some or all of the key references to have been paid less attention? In order to investigate this question we need to know what the "real" content of the article is. Stewart's work is nonetheless an important first step which suggests that characteristics of articles themselves probably have a significant effect on their reception by the broader scientific community.

The "Interests" Theory

In attempting to explain both why individual scientists adopt a particular position and why the community accepts one interpretation over another, some constructivists employ an "interests" theory. Interests can derive from other institutions of society, such as economic, political, or religious interests, or from scientists' professional concerns with matters internal to science (for example, supporting the work of one's students and friends). Interests can also be cognitive, in that each scientist is viewed as having a stake in accepting new ideas which will increase the credibility and importance of her own past contributions and in rejecting ideas which are likely to reduce the credibility or importance of her own past contributions.

Pickering (1984) employs the cognitive interests theory to explain why in high-energy physics the "charm" interpretation of the J-psi particle won out over the "color" interpretation. The claim is that for each scientist there are some data, theories, or models which will support her own work. Because each scientist has invested a substantial amount of time in her career, she will favor new work which fits her cognitive interests. Those ideas which best support the cognitive interests of the most powerful members of the scientific community will win out. But if interests are *cognitive*, it does not seem a convincing argument for the *social* determination of the content of ideas to say that such interests influence the reaction to new contributions. What makes the interests approach unsatisfactory is that it tends to be a tautological and *ex post facto* argument. Those ideas which win out are defined as supporting the interests of the majority of the community. Since the ideas *did* win out, which means that they led to further research, of course it is true that these ideas were useful to other scientists in doing their own work. But for the interests theory to be convincing, constructivists would have to show, first, that the interests of the community could be determined *prior* to a decision concerning two competing ideas and, second, that scientists are not willing to change their views in the light of new contributions which challenge their prior interests.

Galison (1987) provides many convincing examples of scientists who abandoned or changed their interests in the light of empirical data which they saw as contradicting their prior position. Analyzing the experiments on neutral currents discussed above, he asks:

> Is experimentation reducible to such a calculus of interests? As we saw, the E1A team found for some time that there were no neutral currents—they wrote letters saying so, even drafting a paper to that effect. By late 1973 they had a great deal riding on that claim. A consensus that neutral currents did not exist would have vindicated their earlier caution; they would have refuted CERN and denied the Europeans priority. For all these reasons it is stunning to reread Cline's memorandum of 10 December 1973 that began with the simple statement, "At present I do not see how to make this effect go away." With those words Cline gave up his career-long commitment to the nonexistence of neutral currents. "Interest" had to bow to the linked assemblage of ideas and empirical results that rendered the old beliefs untenable, even if they were still "logically possible." (p. 258)

Galison gives another example of the physicist Carl Anderson, forced by data from his own experiments to abandon the theory of his powerful mentor, Robert Millikan: "Anderson could not dismiss the high-energy particles he had observed. Millikan controlled the purse strings at Caltech, his power in the physics community vastly overshadowed Anderson's, and as we have seen, Millikan was almost obsessively attached to his birth cry theory. Certainly Anderson's experiments and interpretations were molded in an environment shaped by Millikan's program. Yet Anderson did eventually dissent, as his own data pointed more and more clearly to the presence of particles too energetic to be of service to Millikan's pet idea" (pp. 258–259).

Is Relativism Necessary?

To what extent is the relativistic perspective assumed by the constructivists either necessary or helpful for their sociological analysis? Their *description* of how ideas move from conjectures to facts is not incompatible with a realist philosophical position. At the time when an idea is first considered it does not make any difference for either relativists or realists whether the idea is really right or wrong, because whether or not there is indeed a right and a wrong, the scientist has no way of knowing what it is. Research in the history and sociology of science suggests that there often are no clear-cut rational methods of immediately determining whether tentative research frontier solutions to problems are right or wrong. Scientists therefore work within a social community, influenced by social processes, to try to create a solution which will be credible or acceptable to other scientists. This description can be accepted by both relativists and realists and was in fact treated in Merton's 1938 monograph ([1938] 1970, p. 219).[17] It is only when we come to *explaining* why some attempts at establishing credibility turn out to be more successful than others that we come to grounds of disagreement between relativists and realists. Realists argue that the content itself has a significant role in determining acceptance; relativists argue that it has little or no role.

Although the relativist assumption may have been useful in stimulating sociologists to look at how scientists go about their work in the laboratory, it is a hindrance to understanding how the work produced in the laboratory is received by the rest of the scientific community. Granted that decisions in the laboratory are not made according to a set of rules which would enable all scientists to make similar decisions. Granted that in making their decisions scientists are influenced by a whole range of social processes, such as the desire to please colleagues or to gain recognition at the expense of colleagues, or by the operation of intellectual authority. But what happens when a piece of work in the laboratory is concluded and a scientific paper is published?

Over the very long run, as Kuhn's classic work emphasizes, ideas which once were the most firmly established part of scientific knowledge can be replaced or even come to be seen as wrong. Whether this is interpreted as meaning that there is no such thing as "truth" or no "out-there-ness," as the relativists would have us believe, or that the ultimate answer to scientific questions is only to be attained in the distant future, as the realists would have us believe, makes very little difference to sociologists studying the behavior of scientists. What should make a difference for sociologists is that scientists *believe* that there is a nature and that some facts conform with it and others do not. As W. I. Thomas said: "If men define situations as real, they are real in their consequences."

If social constructivists take the position that there is no way for the scientific community to establish what is right and what is wrong, and if all knowledge is socially constructed, then one answer to a scientific problem is no better than another; any piece of scientific knowledge could have had some different content. Consider again Guillemin and Schally, who sought to discover the chemical sequence of peptides of the substance known as TRF. They found a particular sequence (Pyro-Glu-His-Pro-NH$_2$), which the scientific community agreed was correct or should be treated as a fact. This fact then became accepted as part of the core and Guillemin and Schally ultimately received the Nobel Prize in medicine. But constructivists would have to argue that it is possible (indeed, quite possible) that some other structure for TRF could have been identified, that this other structure could have been accepted by the scientific community, and that the specialty of neuroendocrinology would have "progressed" to the same degree (or perhaps to an even greater degree) had some other structure been identified. Yet it is unlikely that there is a single neuroendocrinologist in the world who would accept such a conclusion. It is possible for social constructivists to argue that *all* neuroendocrinologists have been socialized to believe in a view of science which is false. Although science does undoubtedly socialize its members into a belief system which is in some sense mythology, a more plausible and parsimonious explanation for the failure of practicing neuroendocrinologists to accept the Latour and Woolgar view is that defining TRF as Pyro-Glu-His-Pro-NH$_2$ works. Accepting the definition enables other scientists to solve scientific problems in ways that fit in with all the rest of the knowledge that they have. Using the Guillemin and Schally chemical sequence for TRF leads to expected and interpretable results, regardless of where the scientists are working or what particular situational contexts and social processes are influencing them. Most, perhaps all, other possible definitions of TRF would have had less utility.

Although in some ultimate sense it is possible that any given solution to a scientific problem can turn out to be wrong, within what Kuhn calls paradigms there are the functional equivalents of right and wrong. Whether or not the structure of TRF turns out in the light of history to have been right, it would not have been possible *within the limits of the current paradigm* to come up with a different finding that both would have been universally accepted by the community (as was TRF) and would have allowed the specialty to progress as it did. Wrong findings could certainly have been proposed, accepted by some members of the community, and have led to some further research. But a wrong finding would not have resulted in the same scientific and applied success which followed the discovery of TRF.

Both further scientific research and applications can be based on incorrect science. Drugs have been prescribed and operations have been performed, for example, on the basis of medical research which we now know to be wrong. Even today knowledge which many people believe to be wrong continues to serve as the basis of applications. Many medical researchers believe that the coronary by-pass operation is effective in prolonging life in only a small minority of cases and may even be of only limited use in relieving pain, yet hundreds of thousands of these operations are performed every year (Horowitz, 1988). I would add, however, that applications based upon erroneous science do not actually accomplish their intended goal. The coronary by-pass operation does not save many lives. In order for the construc-tivist position to be supported it must be shown that applications based upon false knowledge are utilized *and* actually succeed in ac-complishing their manifest function. A successful solution to an ap-plied problem could, of course, be based upon a theory which later turned out to be incorrect or incomplete, as in the AIDS example discussed in Chapter 1. But in this case the solution to the applied problem could not be socially constructed independent of constraints imposed by nature.

Latour and Woolgar occasionally show that they are aware of the problems that relativism introduces. At one point they describe the strategy used by Guillemin, who changed the field of neuroendocrin-ology by insisting upon employing much stricter criteria and methods for determining whether a particular substance was in fact a new and discrete releasing factor. After tightening up the methods, he began to search for the chemical structure of TRF. Latour and Woolgar ([1979] 1986) correctly point out that there was nothing inevitable that caused the specialty to develop in the way that it did: "Because of the success of his strategy, there is a tendency to think of Guillemin's decision as having been the only correct one to make. But the decision to reshape the field was not logically necessary. Even if the decision to pursue the structure of TRF(H) had not been taken, a subfield of releasing factors would still exist. *Of course, only crude or partially puri-fied extracts in short supply would be used,* but all the problems of phys-iology could nevertheless be studied, *if not resolved*" (p. 119, italics added).

In considering the history of any scientific area we should always be aware that in retrospect the way in which the area developed appears inevitable, but that in fact other kinds of development could have been possible. The question is whether another course of devel-opment would have had as much utility in enabling scientists to do their work. The constructivists would have to say that it would. But in the above quotation, Latour and Woolgar recognize that this is unlikely. To say that the problems could have been studied but "not resolved," had some other approach been utilized, is to say that other possible approaches would *not* have been as useful to the scientific community.[18]

At another point in relating the history of the attempt to discover the chemical sequence of TRF Latour and Woolgar note that the ability of the scientists to do their work was limited by the difficulty of obtaining sufficient quantities of hypothalmus to study. They point out that in 1966, three years before the chemical structure was discovered, the scientists might have been forced to abandon the program: "It was then feasible that partially purified fractions would be continued to be used in the study of modes of action, that localization and classical physiology could have continued, and that Guillemin would merely have lost a few years in working up a blind alley. TRF would have attained a status similar to GRF or CRF, each of which refers to some activity in the bioassay, *the precise chemical structure of which had not yet been constructed*" ([1979] 1986, p. 128, italics added). Here, too, Latour and Woolgar are aware that if other possible outcomes of the research program had occurred, then the useful knowledge, the chemical sequence of TRF, would not have been obtained.

Multiple Discoveries

Problems are also posed for the social constructivist view by the existence of multiple discoveries. This problem has been raised in a critique of Knorr-Cetina's work by Freudenthal (1984) and discussed by Zuckerman (1988). If the content of a scientific discovery is a result of idiosyncratic contextual contingencies, how can it be explained that different scientists working in very different contexts frequently make the same scientific discoveries? The implication of multiple discoveries is that discoveries, rather than being socially constructed, are inevitable. It was in part a series of chance factors which enabled Watson and Crick to discover the structure of DNA. But although chance determined *who* made the discovery, would the discovery as accepted by the scientific community have been any different in *content* if someone else had made it? If Merton's (1973, chaps. 16–17) argument that virtually all scientific discoveries are multiples or forestalled multiples is true, this suggests that there are factors beyond local contexts which influence the content of discoveries which are accepted by the community.

Freudenthal (1984) claims that the constructivist interpretation cannot account for cases in which almost identical research is done by different scientists, "because the notion of a scientific work (product) that is almost identical to another cannot be reconciled with the idea that different operations yield different products" (p. 292). The constructivists, recognizing the significance of multiple discoveries, have argued that the research done on multiples is based upon old-fashioned positivist interpretations of science in which ideas are historically reconstructed to make them seem the same when in fact they were very different. Most of this research is based upon historical case studies (Aronson, 1984; Brannigan, 1981; Elkana, 1970). Brannigan, for example, claims that the discoveries of Mendel were not in fact rediscovered in 1900, but that Mendel's work was reinterpreted in order to make it appear similar to the later work. Granted the possible validity of such arguments, it still is evident that many contemporary scientists believe that the solutions they have reached have been independently "discovered" by other scientists (Hagstrom, 1974). Why would scientists who have an interest in claiming sole credit for their contributions be willing to admit that others have independently made the same contribution? Until there is more evidence that all multiple discoveries are simply the artifact of positivist history of science, they remain a potentially troublesome problem for constructivist sociology of science.

Conclusion

Constructivist sociologists of science have no convincing way to explain why some ideas win out in the competition for communal acceptance and admission to the core. Denying the importance of the cognitive content of the contribution as an influence on its reception is equivalent to saying that the evidence introduced at trials has no influence on decisions made by juries. Certainly, as much interesting work in the social sciences has demonstrated (Loftus, 1979), the evidence does not explain all the variance in the outcome of trials. But if we were to try to predict the outcome of trials without having any knowledge of the evidence, we would not have as successful a prediction rate as we would knowing the nature of the evidence. The constructivists say that we should be able to predict which papers will be successful without knowing anything about their cognitive content.

The strategies which sociologists like Latour analyze do have an important effect on a contribution's reception, but strategies do not explain all the variance. Rather we must look at the interaction of social strategies and cognitive content in the evaluation process. "Interests" may also be important in decisions made by individual scientists, but again they do not seem to be able to explain communal decisions without reference to the content of the contributions being evaluated.

The relativist epistemological position of the constructivists is a hindrance to sociologists trying to understand the behavior of scientists. It forces us to deny the importance of evidence from the empirical world and to define all scientists as being deluded about the nature of the enterprise they are engaged in. Relativism may have been necessary and useful in getting sociologists to think about the importance of social processes in the production of knowledge in the natural sciences, but this once-revolutionary approach has now become a hindrance to detailed studies of how social processes interact with cognitive factors in the evaluation of new scientific contributions.

· · 3 · ·

Constructivist Problems in Demonstrating Causality

In considering the research program of the constructivist sociologists of science, we must ask whether they have adequately demonstrated a causal link between social variables and the specific cognitive content of science. I will argue that their studies fall short for two reasons. First, there is a confusion in the work of social constructivists about how a *social* as opposed to a *cognitive* influence should be defined. They frequently refer to decisions made on cognitive or idiosyncratic psychological bases as being social, because the decisions are made in a social environment. But I believe that if at least an *analytic* distinction cannot be made between social and cognitive influences on science, the theory of the social constructivists becomes true by definition: a tautology which adds little to our understanding of science.

Second, I believe that the social constructivists have failed to establish any linkages or connections between the social processes they describe and what I call knowledge *outcomes*. A knowledge outcome is the specific content of a scientific idea such as Einstein's famous equation $E = mc^2$. They succeed in showing that scientists as they carry out their work on a day-to-day basis are heavily influenced by social processes, but they frequently fail to show what influence these processes have on the actual content of scientific knowledge. They are often as "guilty" as the Mertonians of black-boxing the actual cognitive content of science. Because they study social processes on a micro-level as cognitive decisions are being made, they claim that they are showing how these social processes influence the cognitive content of the productions. But they do not give convincing examples of how variation in social independent variables influences knowledge outcomes, and for the most part they do not even attempt to provide such evidence. They succeed admirably in illustrating that the process of doing science is social, but not in demonstrating that the specific content is dependent upon social processes.

Cognitive and Social Influences

All human behavior is conducted in a social context and is therefore influenced by society. But to say that every aspect of human behavior is influenced by society is not to say that the *only* or even the *most important* determinants of any particular behavior are social. For most aspects of human behavior there are nonsocial as well as social causes, as some programmatic statements of constructivists recognize. Bloor (1976), for example, in stating the principles of the "strong program" says: "Naturally there will be other types of causes apart from social ones which will co-operate in bringing about belief" (p. 4). In their writings, however, most constructivists blur the distinction.

Because social and cognitive influences are empirically intertwined does not mean that it makes sense to treat them as analytically inseparable. Suppose that we want to understand why there has been a rapid rise in the proportion of babies born by Cesarean section in the United States. There are many possible causes for this change. It is possible that doctors benefit economically from performing this operation; it is possible that doctors are trying to preserve childbirth as an act controlled by the medical establishment; it is possible that the professional ideology of doctors causes them to make "interventionist" decisions; it is possible that changes in the way malpractice cases are handled cause doctors to perform the operation as part of "defensive medicine"; it is also possible that Cesarean sections may save the lives of babies and reduce the incidence of birth defects. This last reason might more logically be termed medical rather than sociological.[1] If one of the reasons for the increase in the rate of Cesarean sections were medical, most people would consider the sociological explanations to be less interesting than if medical reasons were found not to explain the change. This example gives us some insight into why the constructivists are so insistent in denying cognitive explanations for the development of scientific knowledge. If cognitive explanations hold, interest in the sociological explanations will be reduced.[2] To deny the importance of nonsocial variables, however, makes it impossible to focus on the crucial interaction between social variables and nonsocial variables in influencing specific behavior or outcomes.

The social constructivists fail to make a distinction between social and cognitive influences on the doing of science, and some go so far as to say that this distinction makes no sense (Latour, 1987). The basis of their argument is that science is a communal activity, one which cannot be meaningfully practiced in isolation from others. The fate of any scientific work is in the hands of other people, and work by one scientist can affect the career and work of others. For a scientific production to become a fact requires the recruitment of many allies inside and outside the scientific community. The outside allies, such as funding agencies, are necessary to make the scientific work possible in the first place. Once the work is produced, inside allies, including other scientists who will support the new idea, are necessary for the work to be accepted as a fact. In trying to recruit allies scientists interact with other people both directly and, through the literature, indirectly. This interaction is characterized by the same type of social processes which characterize interaction in other realms of social activity. Therefore, science is just as social as any other activity.

There are few, if any, sociologists who would choose to disagree with this position. And in the case of science, given that the social elements have long been ignored owing to the positivistic bias of those who have studied and written about it, to point out that science is inherently social is an important and useful contribution. But by rejecting even the analytic distinction between social and nonsocial (cognitive) influences on the doing of science, the constructivists impede our ability to specify how sociological variables influence which aspects of science in what ways.

Let us examine again the discussion by Latour and Woolgar ([1979] 1986) of the decision by Guillemin to raise the methodological standards in neuroendocrinology. Guillemin decided that the field could not progress unless it was better able to determine whether a chemical substance was in fact a new and discrete releasing factor. As a result of Guillemin's decision and the willingness of the scientific community to accept the legitimacy of that decision, work which had previously been given credibility lost its credibility: "Data, assays, methods, and claims which might have been acceptable in relation to other goals, were no longer accepted. Whereas Schibuzawa's papers might previously have been accepted as valid, they were subsequently regarded as wrong. That is to say the epistemological qualities of validity or wrongness cannot be separated from sociological notions of decision-making" (p. 121).

The decision by Guillemin to raise the methodological standards is called a "sociological" one. But what is meant by "sociological" in this case? Guillemin believed that progress in the field was being inhibited by the type of methods employed. As Latour and Woolgar say in a sentence prior to this quotation: "In other words, there was no easy shortcut between what was already known and the problem of sequence" (p. 121). Unless all decisions are defined as social because they are made by a person working in a social context, Guillemin's decision could be considered a cognitive one.

Latour and Woolgar seem to be aware of this problem and attempt to bolster their analysis by suggesting that Guillemin's motives were not based upon cognitive criteria. They imply that he made the decision in order to eliminate competition. It is true that the decision by Guillemin to raise the methodological standards had the consequence of mooting the research program of other scientists: "The results of the fresh accumulation of constraints was to put Schreiber out of the race. By increasing the material and intellectual requirements, the number of competitors was reduced" (p. 123). That the consequence of a scientist's decision hurts other scientists, however, does not mean that this motive was primary for the scientist making the decision. And even if direct proof (a scientist's statement, for example) existed that a scientist was motivated in such a way, it would not be particularly informative. We know that scientists, like other humans, want to succeed in their careers and that there is substantial competition among scientists. It would be far more convincing to show that there were *noncognitive* or social reasons why other scientists, including Guillemin's competitors, agreed to go along with Guillemin's decision. Why did the scientific community accept it if they did not perceive the new strategy as having a higher utility for solving the specialty's problems? Latour and Woolgar give us the answer. There were good cognitive reasons for adopting the new strategy: "Previously, it had been possible to embark on physiological research with a semipurified fraction because the research objective was to obtain the physiological effect. When attempting to determine the structure, however, researchers needed absolutely to rely on the accuracy of their bioassays" (p. 124). A new technique was developed which made it easier to distinguish signal from noise. This technique was perceived by other scientists, even competitors of Guillemin, as yielding more useful results; therefore, other researchers went along with his decision. That the decision was made and then accepted in a social context does not mean that it cannot be explained on cognitive grounds.

In Latour's later book (1987) he goes even further in arguing against what he believes to be an arbitrary distinction between science and the society in which it is practiced. He contends that a sociological analysis of science need not refer to the influence of social classes, capitalism, gender, and so on. Not only is science completely social, but the demarcation between science as it is done in the laboratory and the rest of the world is arbitrary. He shows that in order for scientists to create the internal world of the laboratory, where science seems to be done in isolation from the concerns and even from the needs of the rest of society, many allies outside of the laboratory are needed. He gives examples of West, at Data General, who has to work outside of the laboratory to get the resources to continue the development of a new computer, and of "the boss," an eminent scientist who travels all over the world seeking to gain resources for his work and to convince others to do and use science in the ways he wants. He compares West and "the boss" with a Brazilian scientist who wants to develop microchips but receives no social support and is eventually forced to abandon science. Latour argues that if we look only at the people in the laboratory we see science as separate and isolated from the rest of society, but that if we follow West and the boss we see how science depends upon and influences what appear to be external institutions. Science cannot exist without being connected with the rest of society.

> The first lesson to be drawn from these examples looks rather innocuous: technoscience has an inside because it has an outside. There is a positive feedback loop in this innocuous definition: the bigger, the harder, the purer science is inside, *the further outside other scientists have to go.* It is because of this feedback that, if you get inside a laboratory, you see no public relations, no politics, no ethical problems, no class struggle, no lawyers; you see science isolated from society. But this isolation exists only in so far as other scientists are constantly busy recruiting investors, interesting and convincing people. The pure scientists are like helpless nestlings while the adults are busy building the nest and feeding them. (Latour, 1987, p. 156)

Latour is aware, of course, that it is possible to look at the activities outside of the laboratory not as science itself but as auxiliary activities which make science possible. In this case the external activities of science are seen as influencing the foci of scientific attention, as scientists and their patrons negotiate what problems the scientists should work on. The external activities also influence the rate of advance of particular lines of scientific work, as some receive the external support necessary for advancement and others do not. Latour does not give examples of how the "external" activities of scientists influence the actual cognitive content of the work formulated in the laboratory. The utility of lumping everything together has not been demonstrated.

The failure to distinguish conceptually between a social influence and either cognitive or idiosyncratic psychological influences may also be illustrated by Pickering's 1980 article entitled "The Role of Interests in High-Energy Physics: The Choice between Charm and Colour." He describes the experiments done on the J-psi particle by a team led by Samuel Ting at Brookhaven National Laboratory on Long Island and by a team led by Burton Richter at the Stanford Linear Accelerator in California. These researchers were later awarded the Nobel Prize in physics for their experiments. Pickering's goal is to explain why one of two different explanations initially proposed for the experimental observations was accepted by the scientific community. One theory was called "charm" and the other theory "color," and charm quickly came to be the universally accepted explanation of the observation. As noted in Chapter 2, Pickering tries to explain the outcome by using the concept of "interests."[3] What Pickering means by "interests" is simply the interest of a particular scientist in having her own work accepted. The account he presents seems to offer as much support for a positivistic view of science as for a constructivist view. The charm theory was perceived by the physicists to be supported by the data, and the color theory was perceived as being at odds with the data. The latter could only be supported by ad hoc arguments. Further, as Pickering emphasizes, the color theory did not fit in as well with other ideas in the field.

How does this explanation offer us sociological insight into the behavior of the physicists? That scientists are interested in advancing their careers is obvious and has been the subject of research for many years. Thus much of Pickering's analysis seems self-evident: "One can speak of the group or groups having expertise relevant to the articulation of some exemplar as having an 'investment' in that expertise, and, as a corollary, as having an 'interest' in the deployment of their expertise in the articulation of the exemplar. *An 'interest' then, is a particular constructive cognitive orientation towards the field of discourse*" (1980, p. 109, italics added). Despite his use of the term "constructive," even Pickering sees his "interests" as part of the cognitive ideas of the scientists rather than as some social variable. Even if by some stretch of the concept "social" we were to define an interest as such, the account offered by Pickering does not show that interests were more important than cognitive criteria in determining the outcome of the debate over the two theories. For example, he reports the results of experiments which produced a setback for the color theory because they were in contradiction with expectations based on this theory: "It [color] soon acquired a[n] . . . obstacle to acceptance. Early in 1975, detailed experiment on the psi and psi-prime had revealed that photons (quanta or the electromagnetic field) were only to be found infrequently amongst their decay products. This was a considerable setback for the colour model since, straightforwardly articulated, the model predicted that decays involving photons would be the predominant mode" (pp. 120–121).

In discussing the development of consensus in the high-energy physics community on the charm model Pickering concludes: "In more straightforward language . . . it was obvious to all concerned that the charmonium model was right" (p. 123). He does not present any evidence to demonstrate that this conclusion was influenced by social factors or contexts. The analysis does not tell us anything which we could not have learned from reading an account of the discovery written by a traditional internalist historian of science. The sociology is unsuccessfully tacked on.

A similar confusion between cognitive and social influences is found in the Gilbert and Mulkay monograph, *Opening Pandora's Box* (1984). In their chapter on accounting for error they say that the scientists they interviewed always assumed that their scientific opinion was based on empirical evidence but explained the errors of others by "nonscientific" factors. But what is a nonscientific factor? It turns out that this is generally a claim that the other scientist, the one who is wrong, did not understand chemistry (pp. 75–76). It is difficult to see why "failure to comprehend" should be classified as a social influence, as opposed to a cognitive or an idiosyncratic psychological influence, even though both sides in the dispute claim that the other does not understand.

The same problem may be seen in Harvey's (1980) analysis of experimental tests of quantum mechanics. He seeks to explain why there has not been more concern in the physics community with the "hidden variable" theory. This theory, initially proposed by Einstein, contradicts some of the assumptions of quantum mechanics. In discussing some of the "sociological" reasons why physicists have not paid more attention to the hidden variable theory, Harvey mentions variables such as possession of the necessary experimental skills and the criteria by which topics for doctoral dissertations are chosen. It is unclear why such variables are social. Furthermore, in this case, even if such variables are defined as social, they tell us something about the foci of attention of scientists, but nothing about how the specific content of any scientific work is influenced by social variables.

In order to learn whether something has an effect, we must know what that something is. To label everything social, even if everything has some social components, does not further our understanding. The argument should not be over whether social variables influence science, but over exactly what types of social variables influence science in precisely what ways. The distinction between the actual cognitive content of ideas and the doing of science must be kept in mind. The doing of science is, of course, influenced by the micro sociological interaction variables emphasized by the constructivists, but they have failed to demonstrate that the content of any communally accepted ideas is influenced by these variables.

Linkages between Social Processes and Cognitive Knowledge

In order for the constructivist theory to be successfully demonstrated, it must be shown that the form or content of knowledge outcomes (the dependent variable) is influenced by or *caused* by social processes or variables. Although the social constructivists occasionally attempt such analyses, I have not found any which have all the elements needed for a convincing demonstration. There are three difficulties. First, they sometimes show how a specific social process influences the foci of attention or rate of advance, but not the actual content of knowledge outcomes. Second, they sometimes show that the doing of science is a social process, but fail to show how these social processes influence the science being done; the science is essentially black-boxed. Because no causal connection is made between the social processes and the science, they provide an interesting description of how science is done but not a demonstration of their theory. Third, they sometimes show the connection between some social processes in a laboratory and some cognitive decisions made in the doing of science, but fail to show whether or not these differences had any impact on the content of the knowledge outcome.

Foci of Attention and Rate of Advance

First let us look at an example of how the social influences described by the constructivists may help to explain the foci of attention or rate of advance, but not the actual content of scientific knowledge. Consider Latour and Woolgar's ([1979] 1986) account of the reaction of Schally (Guillemin's competitor) to a paper published by Guillemin in 1966. At that time Guillemin had become discouraged and had published a paper which suggested that TRF might not even be a peptide. The work that Schally was then doing was very close to identifying what ultimately turned out to be accepted as the chemical sequence of TRF. Schally's research program was, however, interrupted by his own belief in Guillemin's paper. If Schally had ignored Guillemin's paper and had continued to pursue his own research agenda, he might have won the race and discovered the sequence of TRF more than two years before he and Guillemin discovered it simultaneously (p. 136).

Assuming that Latour and Woolgar are correctly interpreting the reasons for Schally's decision, their account would be a valid example of how noncognitive sociological processes influence cognitive decisions made by scientists. We know that scientists engaged in doing science are heavily influenced by sociological processes, including authority. In this case Guillemin's authority in the field was so great that even his prime competitor went along with what turned out to be a mistaken idea. Perhaps Latour and Woolgar are correct that if Schally had ignored Guillemin's 1966 paper *the* (not *a*) structure of TRF would have been discovered earlier. Latour and Woolgar would have thus illustrated how sociological processes, such as the acceptance of authority, either speed up or, in this case, slow down the rate at which new discoveries can be made. But they do not say that had Schally ignored Guillemin, the content of science, that is, the actual structure of TRF, would have been any different. I would argue that social factors, such as authority, have an influence on the time it takes for discoveries to be made or, once made, accepted by the community, but that they do not have any specific influence on the cognitive content of those discoveries as they are accepted by the scientific community as fact. To state this in another way, Latour and Woolgar have presented qualitative evidence to give credibility to a description of how scientists in their everyday work are influenced by social processes such as authority and the desire to receive social rewards. But they have presented no evidence which enables one to draw a concrete connection between any social process and the specific cognitive content of science. They have not claimed that the chemical sequence of TRF would have turned out differently (and been accepted as fact by the scientific community) had social conditions varied.

Black-Boxing the Content

The second problem found in the analyses presented by constructivists of social interaction in the laboratory is that they frequently black-box the content of the science. For example, consider Latour and Woolgar's ([1979] 1986) analysis based upon observations of two scientists, Wilson and Flower, who were collaborating on research: "Wilson had manufactured these rare and expensive peptides in his own laboratory. So the question for Flower is what quantity of peptides Wilson is willing to provide. The discussion between them thus entails a complex negotiation about what constitutes a legitimate quantity of peptides. Wilson has control over the availability of the substances; Flower has the necessary expertise to determine the amount of these substances" (pp. 156–157).

Latour and Woolgar argue that although the scientists try to phrase their negotiating aims in epistemological terminology, each is really pursuing his own self-interest:

> Given the context of these discussions, it becomes clear that negotiation between Flower and Wilson does not depend solely on their evaluation of the epistemological basis for their work. In other words, although an idealised view of scientific activity might portray participants assessing the importance of a particular investigation for the extension of knowledge, the above excerpts show that entirely different considerations are involved. When, for example, Flower says, "it is very important to do . . . ," it is possible to envisage a range of alternative responses about the relative importance of the uses of peptides. In fact, Wilson's reply ("I will give you the peptides") indicates that Wilson hears Flower's utterance as a request for peptides. Instead of simply asking for them, Flower casts his request in terms of the importance of the investigation. In other words, epistemological or evaluative formulations of scientific activity are being made to do the work of social negotiation. (p. 157)

But in imposing their interactionist perspective in analyzing this conversation, Latour and Woolgar are not giving us an interpretation of how scientists behave which is unbiased by prior preconceptions. By calling the interaction between Wilson and Flower "negotiation" the observers are making the assumption that people define their interests as conflicting. Latour and Woolgar provide no evidence that the participants did not define their interests as coinciding.[4] But suppose that we make the assumption that the two scientists did not see their interests as coinciding. Then we might conclude as do Latour and Woolgar that "nonepistemological" criteria are being used in interaction. But this finding would only be surprising for someone who assumes that all of a scientist's behavior is "rational" behavior dictated by the cognitive rules of doing science.

Most important, the analysis of the exchange between Wilson and Flower fails to show how any aspect of that interaction influenced the content of the scientific work they did. Latour and Woolgar are in fact black-boxing the content. If we assumed that the interaction as interpreted was typical, we would have an empirical generalization which would say that social processes (such as negotiation to attain one's interests) influence the interaction of scientists as they are doing science. We would not know, however, what, if any, linkages existed between these general social processes and the specific content of the scientists' cognitive ideas.

Some of Knorr-Cetina's (1981) examples of scientists negotiating in the laboratory also black-box the science. A scientist she calls Dietrich wanted to perform an experiment using the laboratory of another scientist, Watkins. Dietrich, however, did not want Watkins to be a co-author of the paper which would result from the experiment, and therefore he tried to do the work without letting Watkins know exactly what he had done. Later, because he needed more access to Watkins' equipment, he was forced to show Watkins a copy of a paper he was working on. Watkins then became interested in the work and offered the use of the lab to Dietrich. Knorr-Cetina points out that Watkins maintained control over "his" lab by subverting official rules: "Within the indeterminacy created by Watkins' subversion of the rule, Watkins and Dietrich negotiated the outcome of their interaction with varied success, based upon their changing interests and interpretations. The point with respect to rules is that they seem, in this process, to function more as instruments of negotiation or weapons with varied uses, than as stabilising guidelines for action heeded by the various actors" (p. 45).[5] But we are not told what, if any, connection there is between the negotiation over these rules and any cognitive outcome. The social constructivists describe the social processes going on in the laboratory, but they are unable to link particular social processes with the specific content of any of the science. A theory of how social variables influence specific cognitive knowledge outcomes is lacking.

Social Influences on Knowledge Outcomes

The third problem in the constructivist analysis is that the constructivists show how social processes affect the decisions scientists make while doing science but fail to show what, if any, influence these decisions have on knowledge outcomes. Knorr-Cetina (1981) shows that many of the decisions made in the laboratory she observed resulted from both unplanned choices and local contextual norms rather than from following a set of rational procedures. Scientists will use an apparatus they have lying around rather than construct or purchase a new one. They will decide to take certain measurements because they are easy to do, rather than because they are the most theoretically significant. Decisions will be made on the basis of local tacit knowledge. In the laboratory Knorr-Cetina was studying, measurements made with a particular instrument were taken only once. But one scientist who had been trained at another laboratory believed that single measurements taken on this instrument were not reliable and should, therefore, be repeated and an average used. Knorr-Cetina also reports that various scientists, trained at different institutions, had different ideas about when during a chemical process to take certain measurements.

Knorr-Cetina does not show, however, that these haphazardly made decisions resulted in any difference, even in the local knowledge outcome. No evidence is presented that whether one piece of equipment or another was used had a significant influence on the substance of the cognitive finding. Do the different procedures for taking measurements make any difference? Do these procedural differences lead to major clashes of ideas or are they simply part of the large body of background noise that exists in virtually all scientific research areas? It is almost as if constructivists believe that there is only one way to attain a particular piece of knowledge. If we believe that science is not rational, then it is not surprising to find scientists making casual decisions based upon what is at hand. Nor can it be assumed that these decisions make any significant difference in how knowledge is developed in a research area. In fact there are many examples in the historical literature of different types of experiments or the use of differing experimental methods which yielded the same substantive results. Galison (1987, p. 130) provides a good example in the discovery of the muon. The methods used by Carl Anderson and Seth Nedermeyer at the California Institute of Technology were completely different from those employed by Jabez Curry Street and Edward C. Stevenson at Harvard University. But both groups independently discovered the muon.

Knorr-Cetina's (1981) monograph for the most part ignores the problem of how local knowledge outcomes are evaluated by the community. There are several significant differences between this book and that of Latour and Woolgar. Knorr-Cetina deals only with science as it is done in the laboratory and then written up by the scientists for journal publication. Unlike Latour and Woolgar, she does not describe the construction of any science (such as TRF) which has achieved the status of a scientific fact. Because she does not examine what happens to the products of the scientists she observes after the results are disseminated into the scientific community, she does not tell us what influence, if any, the social processes she describes have on the growth of *communal* scientific knowledge.

For example, Knorr-Cetina (1981, p. 5) describes a scientist trying to fit some data to a particular function. She stresses that because the activity is not a rational, rule-determined procedure, it would have been possible to fit the data with different functions. That there is a great deal of leeway for the interpretation of data in many scientific situations is certainly true. But let us consider the significance of the leeway available to the scientist Knorr-Cetina describes. If the work that this scientist is doing does not lead to publication or leads to publication that is ignored, whether the scientist made one decision or another will be of no significance for understanding the growth of communal knowledge. If the work is published and turns out to be significant, however, it will then be examined by other scientists who are working in other settings, in other social contexts, with other sets of interests. If they do not find the particular function selected to be useful, they will question it and propose other functions. As we move from the informal discussion of work among scientists in the laboratory to the journals to the evaluation process, the significance of local contextual factors is reduced and the extent to which a contribution fits in to the existing body of knowledge becomes more important.

Even in Knorr-Cetina's important analysis of the way in which social negotiation influences the drafting and redrafting of a scientific paper we are unable to determine what, if any, cognitive outcomes may be attributed to the social processes she observes. She emphasizes that all the social negotiation and social processes which led to particular choices or "selections" in the laboratory are left out of the paper, but the reader cannot determine which of these selections had what impact on the final product.

The "Strong Program"

Thus far I have focused on the work of social constructivists who have conducted micro-level studies, either by observation or by qualitative interviews, of particular pieces of science. Another branch of social constructivist sociology of science, the "strong program," has attempted to show that the actual cognitive content of science has been influenced by the more traditionally defined social or political attributes of either the scientists making the discoveries or the contexts in which science is done. The work done in this school is sometimes successful in having both a meaningful cognitive dependent variable and meaningful social independent variables, but like the micro-level studies already discussed, it fails to demonstrate a connection or linkage between them.

Although there is nothing inherently Marxist in the position of the chief adherents of this approach, who include the British sociologists Bloor and Barnes, when they present detailed case studies they tend to emphasize the class interests of scientists. Their work has parallels, in fact, with a famous piece published in 1931 (1971) by the Soviet sociologist Boris Hessen, in which he sought to show that the content of Newton's *Principia* was influenced by Newton's class interests.

Bloor (1982) presents a detailed study of the influence of class interests on the work of Robert Boyle. Following the philosopher of science Mary Hesse, Bloor argues that reality is "indefinitely complex" and because of this "all systems of classification simplify what they portray, and this destroys the possibility of a 'oneness' between knowledge and the world" (p. 278). The connection between what scientists discover and the empirical world is so loose that it leaves room for their class interests or other social factors to mold which aspects of the world they choose to emphasize in their work.

Influenced by Kuhn, Bloor points out that at any given time the scientific community will hold on to or "protect" certain ideas, even in the face of negative evidence. According to Bloor which laws are protected is determined by "their assumed utility for purposes of justification, legitimation and social persuasion" (p. 283). He goes on to explain Boyle's rejection of the idea of Lucretius that matter is animate. Bloor's dependent variable (my term) is a change in the views of the scientific community on matter. An inert or passive conception of matter replaced an active or self-moving conception of matter. To explain this change Bloor looks to the social context. In the early part of the seventeenth century, he argues, social control in England was breaking down. There had been a proliferation of radical religious groups such as the Diggers and Levellers. Boyle's class interests differed sharply from those of these radical sects:

> The Hon. Robert Boyle and the future leaders of the Royal Society had other social goals in mind and possessed very different interests from the sectaries. Their personal fortunes were deeply involved in the quest for stable social forms. In 1646 Boyle had noted with alarm that London "entertains . . . no less than 200 several opinions in points of religion," and wanted to "put a restraint on the spreading impostures of the sectaries, which have made this distracted city their rendevous." Boyle had suffered financially during the Civil War (his Irish estates had been lost), but he had thrown in his lot with the new republic and was now reaping the benefits—benefits that were threatened by the continuing turbulence. (p. 286)

Boyle's desire to protect his economic interests influenced his chemical work: "In the place of this animated and intelligent universe [the ideas of Lucretius] Boyle put the mechanical philosophy, with its inanimate and irrational matter. This was then used to bolster up the social and political policies that he and his circle advocated" (pp. 286–287). Bloor comments: "Collapsing natural hierarchies justified them [the Levellers and other radical sects] in collapsing social hierarchies. To say that matter could organize itself carried the message that men could organize themselves. By contrast, to say that matter was inert and depended on non-material active principles, was to make nature carry the opposite message. Here the world was made to prefigure the dependence of civil society on an involved, active and dominant Anglican Church" (pp. 287–288).

Bloor does not, of course, argue that Boyle and Newton (who is also seen to have been influenced by his class interests) were aware that their work was influenced by their economic interests. Scientists believe that their ideas are influenced by evidence and logic. But the analyst, the sociologist or historian of science, can see the "real" motive force behind the creation of a particular idea: "We know enough of the divergent interests of both sides to explain why all these sources of rational evidence lead to such opposing conclusions. Both groups were arranging the fundamental laws and classifications of their natural knowledge in a way that artfully aligned them with their social goals. The political context was used to construct different pictures of the physical world" (pp. 290–291).

What Bloor has done in this example and what other practitioners of the strong program do is to search for linkages between the substance of a scientist's ideas and the assumed economic, social, or political interests of the scientists. Given the great variety of scientific ideas produced it would probably be possible to find examples of any kind of linkage the analyst was interested in demonstrating. But what has decisively not been demonstrated is a causal connection between the ideas and the interests. How do the interests cause the ideas? Do all scientists with certain class interests develop ideas which justify or support those interests? Or do a higher proportion of scientists with certain class interests develop a particular type of idea than do scientists with other class interests? To my knowledge no data of this type have been produced by the strong program.

This failure to demonstrate a causal linkage between a cognitive idea and a social variable has been criticized both by critics of the social history of science (Garber and Weinstein, 1987) and by other constructivists. Knorr-Cetina criticizes the work of Barnes (1977) by calling it a "congruence" approach:

> The question not answered by this approach is exactly wherein, at what junctures, and how contextual factors such as social interests enter particular knowledge objects . . . Congruence approaches as described above [Barnes] infer that such influences have taken place from perceived similarities between imputed aspects of the content of knowledge objects. They do not specify the causally connected chain of events out of which an object of knowledge emerges congruent with antecedent social interests or with other social acts. (1983, p. 116)

In her response to Gieryn's (1982) critique of relativism-constructivism, Knorr-Cetina argues that it is impossible to use what she calls macroscopic perspectives in order to show how sociological variables such as class interests influence the actual cognitive content of ideas. Unless one employs a micro-level approach and examines in very concrete detail how scientists actually formulate their ideas, one will be unable to demonstrate anything more than "congruence":

> In virtue of the approach they prefer, macroscopically oriented sociology of knowledge invariably appears to end up with congruence models which are based on alleged isomorphisms or similarities between the structure or organization of society and the content of our knowledge. What is needed, however, is not a congruence model but rather a genetic model which demonstrates exactly how and in virtue of which mechanisms societal factors have indeed entered (and are hence reflected in) particular knowledge claims. (1982, p. 332)[6]

Although I do not agree that macrolevel analyses are inherently incapable of showing causality, I certainly agree that the qualitative examples presented by adherents of the strong program have failed to provide the needed evidence.

Underemphasis on Published Science

It is also important to note that the constructivists' emphasis on the importance of science done in the laboratory is accompanied by a corresponding underemphasis on the importance of published science. Many observers of science have pointed out that there is a substantial difference between the ways in which science is actually done in the laboratory and the way in which it is written up for publication (Medawar, 1963; Merton, [1938] 1970). The constructivists have described these differences in detail (Gilbert and Mulkay, 1984; Knorr-Cetina, 1981; Latour, 1980). They sometimes give the impression, however, that the work in the laboratory is the important part of science and that the writing up of these results represents a distortion of the "real" science (Small, 1985).

Consider Knorr-Cetina's (1981) analysis of the extent to which social negotiation influenced the sixteen different drafts of a paper she analyzed. All statements indicating the reasons for particular selections or choices were purged from the final draft of the paper, and as a result it was impossible for other scientists to replicate the results reported in the paper simply by reading it.[7] Knorr-Cetina was surprised to find that this even applied to a co-author of the paper. She quotes from an interview:

Question: "But Watkins is your co-author, he has the paper and must have read the procedure!"
Answer: "Yes, but this doesn't mean that they will be able to do the stuff themselves!" (Knorr-Cetina, 1981, p. 128)

Throughout Knorr-Cetina's analysis of the way in which the paper is written, her language indicates that she is disappointed that what "really" has happened in the laboratory is not reported. She uses the word "perversion" to refer to the scientific paper: "Compared with the work observed in the laboratory, the written paper is, as we have seen, a first complete perversion. Indeed, writing itself is an apt medium for such perversion" (1981, p. 132).[8] A similar reaction occurs in Latour's (1980) paper, in which he points out that there is not a close relationship between what was done in the laboratory and the way in which the results are written up for publication. He is particularly struck that the order in which work was done in the laboratory is switched around for presentation in the paper:

> The draft, however, is covered with corrections and small packages of analogs that are added or deleted. One is added because, said *JR:* "I wanted to draw attention to this one," but it could fit in many other places; three other analogs are eventually crossed out because: "I think that these three analogs published by . . . (a competitor) are really bad and I don't want to embarrass him." Another analog has been shifted from position 10: "Yes, I inverted this analog, and put it at the end because it seemed more logical." . . . Temporal markers invent a temporal framework which is as realistic as that of the fairy tales; it is not written: 'once upon a time', but "the early observations" (. . .) "we then looked," (. . .), "we knew that from then on," etc. As in any other *fiction,* actors are made up that undergo transformations or are supposed to be the authors of various actions. (p. 66, italics added)

Even though Latour later says that there is nothing wrong or dishonest in this process of fictionalizing what went on in the laboratory, the use of the word "fiction," similar to Knorr-Cetina's use of the word "perversion," conveys to the reader a sense of disappointment that the real science is not being reported and that scientific papers hide the way in which science is being done. But if science is viewed as a creative process in which ideas are frequently only loosely bound by particular results in the laboratory, the work of creating a paper from a set of diverse and sometimes confused laboratory results can be seen as a part of science of at least equal significance to the work done in the laboratory. This point is made by Medawar (1963) in a paper cited by Knorr-Cetina (1981).

Almost every sociologist who does empirical work and is interested in relatively broad theoretical or substantive issues knows that there is frequently a large gap between the collection and analysis of data on the one hand and the interpretation of that data on the other. When a sociologist sits down with many tables and empirical analyses before her, there is no one way in which the story as it is presented in a paper, chapter, or a book can be told. Putting the research together in a coherent way is a creative act. A similar point is made by Holmes (1987), who relates how Lavoisier's theory of combustion and his "discovery" of oxygen emerged only as he was writing up his results for publication.[9]

Earlier Work by American Sociologists of Science

Beginning in the early 1970s a great deal of work was done by American sociologists of science on the cognitive content of science. Although these sociologists were not constructivists, their work suffered from some of the same shortcomings, tending to describe the cognitive content of a research area but ignore social variables, analyze social variables but ignore or black-box the cognitive content, or discuss both social variables and the cognitive content but fail to demonstrate any linkage between them.

An example of work which simply describes the cognitive content and contains no information on any social variables would be that done by Sullivan, White, and Barboni (1977) on the specialty of weak interactions in physics. Their paper "The State of a Science: Indicators in the Specialty of Weak Interactions" begins with a brief history of the development of the specialty of weak interactions and emphasizes the significance of the work done by the Nobel Prize winners T. D. Lee and C. N. Yang. The data presented are drawn from a group of 4,691 articles published in the sub-specialty of weak interactions between 1950 and 1972. The authors conducted a content analysis of these articles and classified them along several dimensions. They begin by presenting data on the number of theoretical and experimental articles published in this discipline by year from 1950 through 1972. The data show a relatively sharp increase in the number of theoretical articles and a moderate increase in the number of experimental articles, with peaks in both curves at the time of major discoveries.

Sullivan and his collaborators attempt to explain the difference in the rates of growth of theoretical and experimental work by the technological difficulty of experiments done in this area. The paper contains a wealth of empirical information on numbers of different kinds of papers published in this field, numbers of scientists entering the field over time, and the half-life of references in the papers produced over time in the research area. The paper contains no implicit or explicit references to any social processes or any social causation in the development of the weak interactions specialty.[10]

The second type of problem found in the work of American sociologists of science is the black-boxing of the cognitive content. Mullins and colleagues (1977), for example, use block modeling to show that authors of papers which are highly co-cited actually form a specialty. Their research concludes that people who are co-cited tend to interact socially with one another. The data are not used, however, to address interesting analytic questions such as the extent to which particularistic social ties influence the evaluation of scientific work and the distribution of rewards within the scientific community (see Chapter 7). And the authors make no attempt to consider how social variables influence the development of the ideas which are co-cited.

The third type of problem found in the work of American sociologists, the discussion of both social processes and cognitive content but the failure to demonstrate a linkage, is illustrated by Ian Mitroff's (1974) monograph, *The Subjective Side of Science*. Mitroff demonstrates that the "storybook" version of science held by most laypeople and many scientists is simply not correct. According to the storybook version, scientists are emotionally neutral and detached when evaluating scientific evidence and scientific ideas.[11] Mitroff analyzes a series of in-depth interviews conducted with scientists who studied the moon rocks after the Apollo flights and shows that nothing could be further from the truth. The moon rock scientists were highly committed to their own theories and emotionally involved with the evaluation of both the evidence and the theories. The interviews, quoted in detail, are useful in describing how science is actually done, but they do not demonstrate any causal link between a social variable and specific cognitive ideas. The research is successful in showing that cognitive work is done in a social context and is in general influenced by social processes, but does not explore the ways in which specific ideas are influenced by social factors.

Conclusion

The work done by the social constructivists fails to show how the social processes observed in the laboratory actually influence the specific cognitive content of science, although to prove such influence is their programmatic intent. This group of talented sociologists of science has failed to generate a single example or case study which has all the elements of a convincing demonstration in place. Sometimes they black-box the content of science, as in the case of Latour and Woolgar's analysis of the negotiation process between two scientists. Sometimes they fail to identify adequately a sociological independent variable, as in Pickering's analysis of the debate over the theories of "charm" and "color." The most convincing work in this research program is that of Knorr-Cetina. She is sometimes able to relate specific social processes to specific cognitive outcomes, such as in her analysis of the negotiation that takes place in writing a journal article. But in this work she fails to show whether the social processes she describes had any significant influence on the knowledge outcomes. Would the ideas reported by the scientists have been significantly different if the social context or processes had differed? In Chapter 10 I will discuss what type of sociological study, if any, could successfully demonstrate a connection between the cognitive content of science and one or more social variables.

4

Luck and Getting an NSF Grant

We turn now to quantitative research on consensus and evaluation in science.[1] Before examining the processes which may serve to create consensus, we must first consider the descriptive issue of what level of cognitive consensus in fact exists at the frontiers of science. Such data are crucial in distinguishing between the constructivist approach and my approach, on the one hand, and the approach of traditional positivists, on the other. Both of the former expect to find low levels of consensus. The latter expects to find higher levels of consensus, because science is seen as a unique activity in which objective criteria enable the establishment of consensus. Because my own theory and that of the constructivists both lead to the expectation of low levels of consensus on frontier science, the data in this chapter are not relevant for comparing these two approaches.

I will examine data collected in a study of how grant proposals are reviewed at the National Science Foundation (NSF). These data will show that there is so much disagreement among equally qualified reviewers of a proposal that whether or not a proposal is funded is to a large extent a result of luck in the program director's choice of reviewers. That is, if there is a significant difference of opinion among the population of eligible reviewers in how they will evaluate a proposal, which members of that population are selected will have a large influence on the fate of the proposal.

Although consensus has not been a major topic of research in the sociology of science, there have been occasional empirical studies. Most deal with the differences between the natural and the social sciences; these will be discussed in Chapter 5. The studies which have been done on consensus, though fragmentary, suggest that there may be much less consensus in science than the traditional position has held. Carter (1974) found only a moderate correlation between the priority scores given proposal applications by a National Institutes of Health review panel and later priority scores given the same proposals when submitted for renewal. Small (1974) found a *negative* correlation between the evaluation by referees of articles submitted for publication to chemistry journals and the number of citations the articles received after being published. He also found that the articles which were most cited had received more criticism from referees and that the authors had been required to make more revisions prior to publication. Data summarized by Cicchetti (1991) suggest that there is not a high correlation between the opinions of independent referees reviewing the same articles: that is, knowing the opinion of one referee does not give an observer a much higher probability of predicting the opinion of the second referee than she would have by chance. In addition to the relatively few quantitative studies which have been conducted, dozens of qualitative studies illustrate that there is low consensus on research frontier science (Hull, 1988; Pickering, 1984; Rudwick, 1985; Stewart, 1990).

Such investigations seem to contradict the assumption that we should find high levels of cognitive consensus in the natural sciences. Although constructivists have failed to take cognizance of these studies, they offer support for the constructivist position. If consensus can be determined by comparing a contribution with nature or by the application of a set of rational rules to evaluate the validity of a contribution, why do we find little consensus? There are also cases, however, which seem to contradict the assumptions of the constructivists. These include famous cases such as the Watson and Crick model of DNA, documented by historical qualitative case studies, in which consensus over a new idea emerges quickly and becomes well established.

The case studies of theories such as the Watson and Crick model and the studies of consensus on journal articles and grant proposals seem to lead to contradictory conclusions. But contradiction exists only if we expect to find similar levels of consensus in all parts of scientific knowledge. I have argued that knowledge can be divided into two principal components: the research frontier and the core. The lack of consensus at the frontier in the natural sciences is evidenced by studies of journal and grant refereeing and by many of the case studies conducted by constructivist sociologists and contemporary historians and philosophers of science.

The COSPUP Experiment

In the United States today the federal government is the primary source of funding for basic scientific research. The two most important funding agencies are the National Science Foundation (NSF) and the National Institutes of Health (NIH). From 1975 to 1981 I was employed as a consultant to the Committee on Science and Public Policy (COSPUP) of the National Academy of Sciences (NAS) to conduct a detailed study of how the NSF evaluated proposals submitted by scientists.[2] The NSF employs one form of the peer review system in making research grants.[3] For each grant application, an NSF program director selects a number of scientists, generally four or five, knowledgeable in the relevant subject matter, to act as referees. Each reviewer is sent a copy of the proposal and is asked to evaluate it on the basis of its scientific merit and the ability of the principal investigator. Ability of the principal investigator is generally defined as the quality of his or her recent scientific performance. Each reviewer is asked to make substantive comments and to assign one of five ratings to the proposal: excellent, very good, good, fair, or poor. For the purposes of statistical analysis, numerical values were assigned: excellent = 50, very good = 40, good = 30, fair = 20, and poor = 10.

There were two phases of research on the NSF peer review system. Phase I was aimed at describing how the system worked and finding out whether criticisms that the system was dominated by a particularistic "old boys'" club were justified. My collaborators and I collected data on 1,200 applicants to the NSF in fiscal year 1975. For each of the ten NSF programs studied we selected approximately 120 applicants, half of whom were successful and half unsuccessful.[4] Phase II of the peer review study consisted of a large experiment aimed at finding out if independently selected groups of reviewers would reach the same conclusions about samples of proposals as those reached by reviewers used by the NSF. These studies provide us with the most systematic data collected to date on how new ideas at the research frontier of science are evaluated.

To what extent can we take data collected about research proposals as indicative of the level of cognitive consensus at the research frontier? Research proposals are an important part of work done at the frontier. Without financial support it is frequently impossible to develop new ideas. Research proposals also often involve a discussion of the scientist's past work on the topic. (It is not uncommon for scientists to write proposals for research they have already completed.) It is possible that there might be higher levels of consensus on finished papers than on proposals, but two sets of data suggest that this is not the case. The first are findings from studies of agreement among reviewers of articles submitted to journals. In a comprehensive review of the literature Cicchetti (1991) reports that most studies show Kappa values (the statistic he believes best measures agreement among reviewers) of less than .40, a value which represents what he terms a "poor" level of agreement. He quotes Lazarus (1982) as concluding that only in a small minority of cases do the two referees sent articles by the *Physical Review Letters,* one of the two most prestigious physics journals in the world, agree on acceptance or rejection. All available evidence suggests that there are substantial levels of disagreement on the publishability of completed articles. The second set of data regarding the level of agreement on scientists' past contributions will be presented below. Both these sets of data suggest that the level of consensus on research proposals will not be significantly different from the level of consensus on other work produced at the frontier. This conclusion is also supported by the many qualitative case studies which show high levels of disagreement on published work at the research frontier.

The data of Phase II of the peer review study are the most relevant for investigating the degree of consensus that exists at the research frontier of science. In the spring of 1977, the NSF provided COSPUP with 150 proposals from fiscal year 1976—50 each from the programs in chemical dynamics, economics, and solid-state physics— upon which decisions had been recently made; half the proposals in each program had been funded, and half had been declined. To select new reviewers we used a panel of from ten to eighteen experts in each of these fields, most of them members of the NAS. Each proposal was sent to two members of this panel, each of whom selected six or more reviewers for it. This gave us a list of approximately twelve new reviewers for each proposal.[5]

The proposals were then sent out to be reviewed by the experimental reviewers. The proposals were reviewed under two conditions. Under the first condition (nonblinded) no attempt was made to conceal the identity of the principal investigator(s), and the reviewers were asked to evaluate the proposals using the criteria employed by the NSF. The NAS letter discussing the criteria of evaluation was identical to the one used by the NSF. In the second condition (blinded) we attempted to conceal the identity of the principal investigator(s) by removing title pages containing the names of principal investigators and their institutions; removing lists of references, budget, and any descriptions of facilities that might identify the institutions at which the research was to be conducted; and omitting any direct references to past work of principal investigators and any other remarks that would obviously lead to identification.[6] The experimental reviewers were randomly assigned to review either the blinded or the nonblinded proposals. Reviewers of blinded proposals were asked to identify authors of proposals if they could. Whether or not they attempted this, reviewers of blinded proposals were instructed to evaluate the proposal strictly in terms of the content of the science contained in it. They were asked to ignore the investigator's past track record if they thought they knew the principal investigator's identity.

The data from this experiment enable us to answer three important questions. First, to what extent would two independent groups of reviewers reach the same conclusions about which proposals merited funding? Given that the two groups of reviewers (NSF reviewers and COSPUP reviewers) were selected independently, if they reached essentially similar conclusions this would suggest that even at the research frontier of science we could expect to find a relatively high level of agreement. Such a finding would cast serious doubt upon some of the basic assumptions of the social constructivist position and bring into question the conceptual distinction I have made between the core and the frontier. Second, assuming that we did find some lack of agreement between the two independent sets of referees, to what extent would these discrepancies be the result of either bias in the way in which the NSF and/or COSPUP selected the reviewers or high levels of cognitive dissensus? Both the social constructivist position and the conceptualization of the difference between the research frontier and the core would lead one to expect to find discrepancies based upon lack of cognitive consensus. Finally, to what extent were the opinions of reviewers influenced by knowledge of the identity of the author(s) of the proposals? This question is primarily relevant for an analysis of the social processes which influence evaluation of work at the frontier, to be discussed in Chapter 6.

In the experiment we were able to measure the extent to which there was consensus among groups of independently selected reviewers. Would the NSF have reached similar funding decisions if the COSPUP reviewers had been used to make the decision? As already mentioned, in order to conduct this experiment it was necessary to select new reviewers who would be qualified to review each of the 150 proposals included in the experiment.[7] Because the data suggested that the pool of scientists available to evaluate a given proposal was generally considerably larger than the group used by the NSF, we wanted to see how the program director's choice from among the group of eligibles influenced the funding decision reached. Would independently selected panels of experts reach the same conclusions as had the NSF peer reviewers? To find out we compared the evaluations made by the NSF-selected reviewers with those made by the COSPUP-selected reviewers.

First, I shall compare the NSF reviewers with the COSPUP non-blinded reviewers. (The NSF reviewing process does not "blind" proposals.) In general, the COSPUP nonblinded reviewers tended to give slightly lower scores than did the NSF reviewers (see Table 4.1). (The experimental reviewers may have been slightly harsher in their evaluations because they knew their ratings would have no effect on the careers of the applicants.) For example, for the fifty chemical dynamics proposals, the grand mean of the NSF reviewers' ratings was about 38 on the 10-to-50 scale, that of the COSPUP reviewers about 35. The correlations between the mean NSF rating and the mean COSPUP rating for each proposal are moderately high. Proposals that are rated high by NSF reviewers tend also to be rated high by the independent sample of reviewers used by COSPUP. The match, however, is less than perfect.

The experimental data allowed us to ask how many funding deci-

Table 4.1 Correlation of mean ratings of NSF reviewers and COSPUP reviewers on grant applications ($N = 50$)

Program	Mean Ratings		Correlation Coefficient
	NSF	COSPUP	
Chemical dynamics	37.7 (±5.85)	35.0 (±6.45)	0.595
Economics	33.6 (±9.76)	31.5 (±9.28)	0.659
Solid-state physics	38.2 (±6.15)	35.5 (±6.33)	0.623

Source: S. Cole, J. R. Cole, and Simon, 1981, p. 882, Table 1.

Note: Numerical values were assigned to the original ratings as follows: excellent = 50, very good = 40, good = 30, fair = 20, and poor = 10. Figures in parentheses are standard deviations.

sions would be reversed if they were determined by the COSPUP ratings rather than by the procedures followed by NSF? We placed the proposals in each program in rank order according to the NSF mean ratings and the mean ratings of the COSPUP nonblinded reviewers (see Figures 4.1, 4.2, and 4.3; half-integer ranks are the result of ties). Because the mean ratings generally determine which proposals are funded, and half were funded and half declined, decisions on proposals that were ranked in one half of the range of scores by NSF reviewers and in the other half by COSPUP reviewers would have been reversed by the COSPUP ratings. There were differences of that degree in the ratings of approximately a quarter of the proposals.

The NSF is faced, of course, with a zero-one decision rule: to fund or not to fund a proposal.[8] It follows that proposals with mean rankings that are fairly close, or virtually identical, may fall on opposite sides of the dividing line. It was almost inevitable, therefore, that we would find some reversals. In determining what should be classified as a reversal, we rank-ordered the proposals according to their mean COSPUP ratings and assumed that those with the top twenty-five scores would be funded and the bottom twenty-five declined. We then

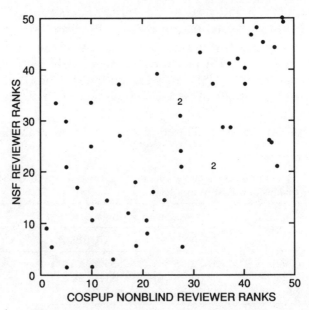

Figure 4.1. Rank order of chemical dynamics proposals as reviewed by NSF and COSPUP nonblind reviewers ($N = 50$). From S. Cole, J. R. Cole, and Simon, 1981, p. 883, Figure 1.

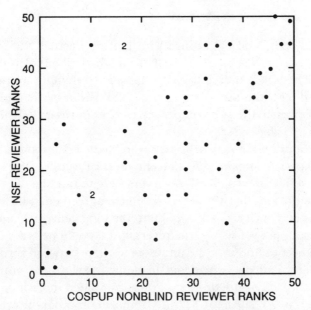

Figure 4.2. Rank order of economics proposals as reviewed by NSF and COSPUP nonblind reviewers (*N* = 50). From S. Cole, J. R. Cole, and Simon, 1981, p. 883, Figure 1.

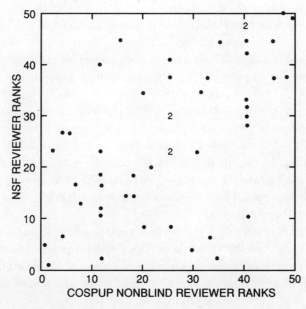

Figure 4.3. Rank order of solid-state physics proposals as reviewed by NSF and COSPUP nonblind reviewers (*N* = 50). From S. Cole, J. R. Cole, and Simon, 1981, p. 883, Figure 1.

compared the COSPUP ratings with both the actual NSF decision and the decision NSF would have made if it had relied solely on mean ratings. The NSF funding decision was highly, though not perfectly, correlated with the mean ratings that NSF reviewers had given the proposals; hence the two comparisons yield approximately the same results (see Table 4.2).[9]

If decisions on the fifty proposals were made by flipping a coin, we would expect to obtain a 50 percent reversal rate, on the average. Correlatively, if the COSPUP reviewers were to rate the fifty proposals in such a way that there was complete agreement with the NSF reviewers on which proposals fell into the top twenty-five and which into the bottom twenty-five, the reversal rate would be zero. Thus we would expect to find a reversal rate somewhere between zero and 50 percent and it in fact turns out to be between 24 percent and 30 percent for each of the three programs computed in each of the two different ways. That is, for 12 to 15 of the 50 proposals in a program, COSPUP reviews would have led to a "decision" different from that of NSF reviews.

We would expect to find some reversals around the cutting point, for example, to find that a proposal ranked twenty-fourth by NSF was ranked twenty-sixth or twenty-seventh by COSPUP. To examine the extent to which reversals were common not only at the midpoint but at a distance from it, we calculated the reversal rates within quintiles (see Table 4.2).[10] In chemical dynamics and solid-state physics we find, as expected, the highest reversal rate in the middle quintile. A 50 percent reversal rate for this quintile would not be surprising. In chemical dynamics it is 60 percent in both comparisons and in solid-state physics 49 percent and 43 percent. In economics, by contrast, we find higher reversal rates in the second and fourth quintiles than in the third. And in all three programs there are proposals that were rated in the first quintile by NSF reviewers that would not have been funded had the decision depended on the appraisals of the COSPUP reviewers.

Several explanations for the reversals are possible. If the two sets of reviewers used different criteria in appraising proposals, the outcome could have differed significantly, creating reversals—for example, if one group of reviewers based its ratings strictly on evaluations of the proposal and the other primarily on the past track record of the applicant. Given that the NSF reviewers and the COSPUP nonblinded reviewers were given identical instructions about the criteria, it is unlikely that there were systematic differences of this kind.

Table 4.2 Reversals in three programs: Comparing NSF ratings with COSPUP nonblinded ratings

Quintile based on COSPUP rating	Percentage of reversals							
	Chemical dynamics		Economics		Solid-state physics		Three fields combined	
	NSF mean rating	NSF decision	NSF mean rating	NSF decision	NSF mean rating	NSF decision	NSF mean rating	NSF decision
1	26	26	20	5	23	16	23	16
2	24	24	45	45	22	24	30	31
3	60	60	30	28	49	43	46	44
4	20	20	45	42	34	29	33	30
5	20	20	0	0	6	11	9	10
Overall	30	30	28	24	27	25	28	26
N cases	50	50	50	50	50	50	150	150

Source: Adapted from S. Cole, J. R. Cole, and Simon, 1981, p. 883, Table 2.

Another possible procedural cause of reversals might obtain if NSF and COSPUP selected different types of reviewers. Reviewer differences rather than proposal differences could then result in reversals. Because a comparison of the characteristics of the two groups of reviewers showed few differences, however, it is likely that they were drawn from the same population.

Assuming that reversals did not result from the procedures employed in the experiment, we are left with two possible substantive explanations. Reversals could result from bias in the way in which the reviewers were selected by either the NSF program director or the COSPUP experiment. If, for example, the NSF program director purposely selected reviewers who would give unrepresentative negative or positive ratings to a proposal, a reversal could result. Or reversals could have resulted from disagreements among fairly selected reviewers using the same criteria. If there were substantial dissensus in the population of eligible reviewers of a given proposal, then it would be possible for equally qualified and unbiased groups of reviewers using the same criteria to differ in the mean rating.

Consider a hypothetical proposal for which there is a population of approximately 100 eligible reviewers. If all 100 were in complete agreement about its merits, then any sample of four or five selected at random from the 100 would agree among themselves, and two independently selected samples would not reach different conclusions. If the eligible reviewers were in substantial disagreement about the proposal, however, two randomly selected samples could yield different mean ratings and possibly lead to different outcomes for the proposal. The data indicate that the reversals in this experiment were a result of such disagreement. The results of an analysis-of-variance model (see the Appendix) showed that there were no significant differences in the way in which the NSF and COSPUP reviewers evaluated the proposals for any reason. We were therefore able to conclude that there was no evidence of any bias in the way in which either the NSF or COSPUP selected the reviewers.

To explain the reversals, then, two sources of variance must be examined: differences among the proposals and differences among the reviewers of a given proposal. In the two physical sciences the variance among reviewers of the same proposal is approximately twice as large as the variance among the proposal means; in economics the reviewer variances are about 50 percent larger than the proposal variance. If the pooled proposal mean (the mean of both sets of ratings in each comparison) is taken as a rough indicator of the quality of the application, the variation in quality among the fifty proposals is small compared with the variation in ratings among reviewers of the same proposal. We treated the reviewer variances as rough indicators of disagreement among reviewers. In all three fields there is a substantial amount of such disagreement.[11] It is the combination of relatively small differences in proposal means and relatively large reviewer variation that creates the conditions for reversals. We concluded that the reversal rate could be attributed to low levels of consensus among the population of eligible reviewers.

When we compared the ratings obtained from the blinded reviews with the two other sets, we found a moderate correlation between the means of the blinded reviewers and the means obtained by the NSF (see Table 4.3). We compared the COSPUP blinded reviewers with the NSF reviewers to ascertain how many reversals there would have been if the decision had been made on the basis of the opinions of the blinded reviewers (see Table 4.4). The reversal rates, computed in the same manner, are similar to those reported in Table 4.2. The overall percentage of reversals is 33 (compared with 27 percent before). When the COSPUP blinded reviews were compared with the NSF funding decisions, the reversal rate was 33 percent (compared with 25 percent previously).

One should expect higher reversal rates when blinded reviews are compared with NSF reviews, because whereas the nonblinded reviewers were asked to use the same criteria as the NSF reviewers, the blinded reviewers were asked to use a different set. They were instructed to evaluate the proposal strictly on its content even if they knew who the principal investigator was. We conducted a components-of-variance analysis in which the blinded and nonblinded COSPUP reviews were compared (see the Appendix). This analysis led to the conclusion that it is reviewer disagreement, rather than

Table 4.3 Correlations of ratings by three sets of reviewers (B = blinded; NB = nonblinded)

Correlation between	Chemical dynamics	Economics	Solid-state physics
COSPUP (B) and NSF	.42	.55	.53
COSPUP (B) and COSPUP (NB)	.58	.63	.53
COSPUP (NB) and NSF	.60	.66	.62

Table 4.4 Reversals in three programs: Comparing NSF ratings with COSPUP blinded ratings

| | | | | Percentage of reversals | | | | |
| Quintile based on COSPUP blinded rating | Chemical dynamics | | Economics | | Solid-state physics | | Three fields combined | |
	NSF mean rating	NSF decision	NSF mean rating	NSF decision	NSF mean rating	NSF decision	NSF mean rating	NSF decision
1	30	30	15	10	20	20	22	20
2	40	50	53	46	38	38	44	45
3	52	52	62	64	22	22	45	46
4	30	30	45	50	20	30	32	37
5	10	20	25	10	40	30	25	20
Overall	32	36	40	36	28	28	33	33
N cases	50	50	50	50	50	50	150	150

blinding, which accounts for nearly all the reversals in a comparison of blind and nonblind reviews. These data reinforce the conclusion that reversals are a result of cognitive disagreement among members of a research area.

The substantial disagreement among reviewers of the same proposals in the experiment can also be illustrated by a simple one-way analysis of variance (see Table 4.5). About half of all the variance in ratings resulted from disagreement among reviewers of the same proposals. I was able to replicate this one-way analysis of variance on the ten programs for which we collected data in Phase I of the research. For each of the programs we had data on a random sample of 100 proposals (see Table 4.6). The results from Phase I support the conclusion that there is substantial disagreement among the several reviewers of a given proposal. The one-way analysis of variance divides the variation in peer review ratings into that which could be attributed to variation among the applicants (or proposals) and that which could be attributed to variation between reviewers of the same proposal. About half of all the variance in ratings is seen to result from disagreement among reviewers of the same proposals. The within-proposal variance accounted for 35 to 63 percent of the total variance in the ten programs.

Another way of conceptualizing the relatively low consensus among reviewers of the same proposal is to think of placing all the NSF reviews of the fifty Phase II chemical dynamics proposals in a hat and drawing two out at random. If we did this a large number of times we would find, on average, an expected absolute difference in the ratings of 9.78 points (the difference, for example, between a "good"

Table 4.5 Percentage of total variance in reviewers' ratings accounted for by differences among reviewers of individual proposals

	Phase II		
Program	NSF reviewers	COSPUP nonblinded reviewers	COSPUP blinded reviewers
Chemical dynamics	60 (242)	53 (213)	63 (212)
Economics	43 (163)	47 (182)	48 (199)
Solid-state physics	51 (192)	49 (190)	51 (203)

Source: Adapted from J. R. Cole and S. Cole, 1985, p. 39, Table 3.4.
Note: The number in parentheses is the total number of reviewers.

Table 4.6 Percentage of total variance in reviewers' ratings accounted for by differences among reviewers of individual proposals

Program	Phase I	
Algebra	45	(315)
Anthropology	43	(227)
Biochemistry	49	(423)
Chemical dynamics	45	(392)
Ecology	54	(402)
Economics	35	(324)
Fluid mechanics	43	(335)
Geophysics	57	(491)
Meteorology	63	(562)
Solid-state physics	55	(498)

Source: Adapted from J. R. Cole and S. Cole, 1985, p. 39, Table 3.5.
Note: The number in parentheses is the total number of reviewers.

and a "very good" rating). Now, if we placed all the reviews of a single proposal in the hat and drew out two, we would find, on average after multiple trials, an expected absolute difference of 8.45 points.[12]

After the results of the experiment were obtained, some members of COSPUP were surprised at the low level of consensus they exhibited and the correspondingly high level of chance in determining which proposals would be funded. It was hypothesized that perhaps there would be more consensus on the past track records of the principal investigators, and that perhaps chance would be less significant in peer review if a system could be devised which would put greater emphasis on this criterion. To find out the level of consensus on past track records my collaborators and I collected data on the assessments of the past research contributions made by the applicants whose proposals were under review during Phase II of the study.

A single-page, two-item questionnaire was sent to the reviewers who had, approximately a year earlier, appraised the 150 research proposals in Phase II. A total of 1,598 questionnaires were sent to reviewers in chemical dynamics, solid-state physics, and economics.[13] Of these, 1,290, or 81 percent, were returned after two mailings.[14] Two questions appeared on the questionnaire. We asked scientists to assess the principal investigator's contribution through his or her published work over the past ten years both to the entire field (question 1) and to the specific research area or specialty (question 2). Five substantive categories made up the rating scale:

Is among the few most important contributors to the field [specialty].

Has made very important contributions to the field [specialty].

Has made above average contributions to the field [specialty].

Has made average contributions to the field [specialty].

Has made below average contributions to the field [specialty].

This is not an interval scale, but it allows for an ordering akin to the adjectival appraisals used by NSF for proposal evaluations, from "excellent" to "poor." Two additional categories allowed the reviewers to indicate that they were not familiar with the principal investigator's work but had heard of him/her or that they had never heard of the principal investigator.

There is a potential problem with this measurement of reputation. The scientists who produced assessments of the reputations were the same scientists, or at least were drawn from the same sample of scientists, who produced evaluations of proposals. More significant, many of the reviewers had reviewed the proposal a year before, and this prior experience could have influenced their judgment of reputation. Assuming that the strain toward consistency is great, reviewers might have either recalled their earlier proposal ratings or looked up the ratings before judging reputations. Although this possibility remains and could only be definitively handled by obtaining appraisals from an entirely independent sample of reviewers, the data suggest that such confounding is unlikely to have occurred. The evidence for this derives from the availability of the NSF ratings of proposals, which were all by raters different from those used in our experiment. In that sense, they represent an independent sample. It turns out that the mean ratings that the NSF reviewers made of the proposals were just as highly correlated with the reputational measures as were the mean proposal ratings derived from COSPUP nonblinded reviewers. Because reputational ratings were not made by NSF raters, the ratings could not be the result of an attempt by the reviewers to be consistent in their proposal and reputational evaluations.

There is further evidence that suggests that the reputational measures are not confounded by the prior evaluation of the proposal. If a respondent's judgment of the reputation of a scientist had been influenced by his previous reading of the proposal, then the differentials between the average peer review ratings produced by the NSF reviewers and by the COSPUP reviewers should be correlated with the reputational ratings that they produce. That is, the proposals to which COSPUP gave higher ratings should have higher reputational scores. To test this idea, we examined the relationship between the difference in the mean NSF proposal ratings and the mean ratings for the nonblind COSPUP proposals, and the mean reputational score received by the proposal.

We examined the zero-order correlation coefficients between the differences in mean ratings and the mean reputational score for the three programs studied in Phase II. There were no substantive or statistically significant correlations between these differences in mean ratings and the reputational scores received by the authors of the proposals. The results are similar whether we relate the difference scores to appraisals of contributions to the discipline as a whole (question 1), to the particular specialty of the principal investigator (question 2), or to the visibility of the principal investigator (that is, whether the evaluator had heard of the principal investigator). None of the correlations was statistically significant. We therefore concluded that the reputational ratings were not significantly influenced by the fact that the raters had read the proposal a year earlier.

The results obtained from these questionnaires were analyzed in the same way as those obtained for peer review proposal ratings. An analysis of variance of the evaluations excluded those cases in which the evaluator did not make an evaluation of the principal investigator's contribution. In this one-way analysis of variance we are interested in the amount of total variation in "reputational standing" that could be accounted for by disagreements among evaluators of the same individual and, correlatively, the proportion that could be accounted for by differences among individuals.

Consider the level of consensus about the reputational standing of scientists whose proposals were reviewed in the Phase II experiment (see Table 4.7). In chemical dynamics, when we examine the amount of the total variance in the ratings of scientists' contribution to their field over the past decade (question 1), we find that 50 percent of the variance is a result of disagreements among evaluators about the individual scientists' reputation. For the more highly specified question about the contribution to the scientists' particular research specialty, 45 percent of the total variance is accounted for by these disagreements. These figures are very similar to those obtained when we examined the level of consensus among raters of NSF and COSPUP research proposals (see Table 4.5).

Table 4.7   Percentage of total variance in ratings of scientific reputation accounted for by differences among evaluators of individual investigators

	Evaluation of principal investigators'			
Field	Contribution to field		Contribution to specialty	
Chemical dynamics	50	(285)	45	(263)
Economics	44	(245)	47	(219)
Solid-state physics	50	(234)	45	(202)

Source: Adapted from J. R. Cole and S. Cole, 1985, p. 45, Table 3.6.
Note: The number in parentheses is the total number of evaluators.

The story is much the same in both economics and solid-state physics. The level of consensus among these peer reviewers about the importance of the past contributions of the research scientists is roughly equal to their level of consensus about the merits of scientific proposals. Approximately half of the total variance in reputational ratings can be accounted for by disagreements among reviewers about the importance of the past research contributions made by the particular principal investigator.

These findings on consensus in reputations can be usefully coupled with the substantial correlation between reviewer evaluations of a proposal and their evaluation of the reputational standing of the applicant. (The average correlation in the three fields was in the range of .50.) There is some consistency within individual reviewers in their evaluations of various aspects of an applicant's work, but there is substantial inconsistency among the reviewers both about the merits of proposals and about scientific contributions made in the past by the principal investigator.

All of the research conducted on peer review at the NSF led to the conclusion that the funding of a specific proposal submitted to the NSF is to a significant extent dependent on the applicant's "luck" in the program director's choice of reviewers. This conclusion should not be interpreted as meaning either that the entire process is random or that each individual reviewer is evaluating the proposal in a random way. In order to clarify the way in which the luck of the draw works, we must look at the sources of reviewer disagreement.

Some of the observed differences among scores given to the same proposal by different reviewers is undoubtedly an artifact of what anthropologists refer to as intersubjectivity. That is, there may be two reviewers who translate their substantively identical opinions differently; reviewer A's opinion is expressed as an "excellent" and reviewer B's as a "very good."

The great majority of reviewer disagreement observed is probably a result of real and legitimate differences of opinion among experts about what good science is or should be. We base this conclusion on the qualitative comments reviewers made both on the proposals studied for the COSPUP experiment and on those studied in the first phase of the peer review study. Contrary to a widely held belief that science is characterized by agreement about what is good work, who is doing good work, and what are promising lines of inquiry, this research indicates that concerning work at the research frontier there is substantial disagreement in all scientific fields.

Conclusion

There has been a division of opinion among scholars and scientists in their reactions to the results of the NSF peer review study. Most have expressed surprise at the level of disagreement and at the extent of the role of chance resulting from that disagreement. But some, including the University of Chicago statistician William Kruskal, Jr., have expressed surprise at the "high" level of consensus: "It is a common observation that experts disagree, and that careful objective studies of expert judgment typically find them disagreeing more than had been expected and more than the experts themselves find comfortable. Consider wine-tasting, economic prediction, demographic projection, medical diagnosis, judgment in beauty contests, diving contests, and cattle shows, and even Supreme Court decisions. Variability is usually considerable, the more so in close cases."[15]

Kruskal is certainly correct that high levels of disagreement exist in all of the evaluation situations he mentions. The problem with dissensus in peer review and in other scientific appraisal systems is not, of course, that scientists disagree. The problem lies in the mythology that they do not or should not disagree. Some observers of science attempt to perpetuate the idea that science, at least natural science, is marked by great consensus. I suspect that few scientists would like to think of peer review of NSF proposals or refereeing of scientific papers at leading journals as akin to the subjectivity in beauty contests or wine tasting. In science there is supposed to be an objective standard to which we can compare a contribution and measure its significance. Objective standards in this sense do not exist for human beauty or for the taste of wine. Furthermore, I suspect that the majority of scientists do not think of decision making at the NSF as influenced significantly by the personal values of the judges—at least not as much as might be seen in an abortion case before the Supreme Court. What makes these data from the peer review study significant is that they are the best data we have on the level of cognitive consensus that exists on research frontier science. These data indicate that levels of consensus are low.

5

Consensus in the Natural
and Social Sciences

Scientific knowledge is not simply the aggregate of the intellectual output of individuals or groups of individuals.[1] What we refer to as science is communally held knowledge; therefore the productions of the laboratory, in order to become scientific knowledge, must reach the attention of other scientists, be evaluated by them, and be incorporated into that body of knowledge which is the collective property of the community. Ziman (1968) makes this point well in his book *Public Knowledge:* "Science is not merely *published* knowledge or information . . . Scientific knowledge is more than this. Its facts and theories must survive a period of critical study and testing by other competent and disinterested individuals, and must have been found so persuasive that they are almost universally accepted. The objective of Science is not just to acquire information nor to utter all noncontradictory notions; its goal is a *consensus* of rational opinion over the widest possible field" (p. 9).

The research analyzed in this and the following two chapters focuses on how scientific work is evaluated. The data from the NSF peer review experiment reported in Chapter 4 provided systematic empirical support for the conclusion that the level of consensus is relatively low at the research frontier, but it did not tell us how the scientists actually made their evaluations of the research proposals. What are the social processes through which ideas at the research frontier are evaluated, and how do some contributions come to be accepted as facts whereas others are ignored or rejected?

As I pointed out in Chapter 1, positivists take the answer to this question for granted. In their view the primary criterion by which new ideas are evaluated is the extent to which they match nature; this determination is made by following a set of agreed-upon rules. Constructivists reject this interpretation because they see nature as created by scientists rather than serving as the arbiter of scientific consensus. Thus far, however, as we saw in Chapters 2 and 3, the constructivists have failed to specify how social variables determine not only the specific content of scientific ideas but which among these ideas are accepted as both important and true.

Evaluation of others' work is a constant part of the activity which scientists engage in on a day-to-day basis.[2] We know so little about the evaluation process that our discussion of the variables involved must of necessity be highly speculative. At least three categories of variables exist: the attributes of the authors of scientific contributions, the process of vesting authority in intellectual elites, and the attributes of the contribution itself.

The first general type of variable possibly influencing the evaluation of a scientific contribution is the social characteristics of the scientist making the contribution. Holding the content of the contribution constant, will the work of a certain type of scientist be more likely to be favorably evaluated and thus move toward the core? A great deal of work has been done on this topic, but it has not focused specifically on how the operation of the reward system influences the development of consensus. Chapters 6 and 7 will examine this question in detail.

The second type of variable involves social processes such as the operation of intellectual authority. Analytically we might even divide the problem of evaluation into two sub-problems: How do authorities determine what contributions deserve their attention? How is the influence of authorities spread through the wider scientific community? Authorities are important because the scientific community does not consist of an atomized set of isolated scientists who decide for themselves whether or not each contribution is important or true. Scientists' opinions about the work of others are only in part based upon a direct and independent reading of that work. Most scientists do not have the time or the inclination to read more than that work which is most directly related to their own research. Even if other work is read, many readers do not have the technical skills or experience to make an independent evaluation of its importance. Most scientists, therefore, are influenced in their opinions of others' work by the opinions of respected colleagues who serve as intellectual authorities. The operation of authority is a crucial social aspect of the evaluation system. This question will be further analyzed in Chapter 8.

Third, we must consider the variable of the attributes of the work itself.[3] To determine whether there are *general* attributes of scientific contributions which influence the way in which they are evaluated is a very difficult task, one which will require the collaboration of historians, philosophers, and sociologists of science over a long period of time. It is imperative that any studies of this variety should be empirical, rather than following the normative approach taken in the past by traditional philosophers of science. In this book I consider only one small aspect of the question: the macro-level problem of how the subject matter of the discipline might influence the ease of reaching evaluative consensus. Is it easier to reach consensus on work in the natural sciences than on that done in the social sciences? The nature of the core in the natural and the social sciences differs significantly. The core of the social sciences is smaller and more diffuse. In the natural sciences there is agreement on specific and concrete theories, such as the Watson and Crick model of DNA. In the social sciences there is agreement that certain general orientations are important (such as that of Max Weber in sociology), but little agreement on specific interpretations or applications of those general orientations. Why do we find these differences? Does nature speak more clearly to natural scientists than to social scientists? If so, we would expect to find differences between the natural and social sciences not only in the core but also at the frontier. Indeed, it is widely believed that the natural sciences do exhibit a higher level of consensus than do the social sciences.

I will present data from a series of empirical studies, all of which suggest that at the research frontier there are not significant differences in consensus between the natural and the social sciences. These studies lead us to conclude that broad variables such as the subject matter of research areas, their level of development, and the content of their cores do not influence the ease of reaching consensus on frontier science. Other data from a preliminary study, however, do suggest that there is more consensus on core knowledge in the natural than in the social sciences.

A significant methodological problem should be noted: What should the proper unit of analysis be for comparative studies of consensus? For most of the studies which I will consider, the unit of analysis has been the scientific field. But in many cases the scientific field may be more meaningful as a sociological entity than as an intellectual one. Should we compare broad areas such as the traditional scientific fields (physics, chemistry, psychology, and so on), research published in a particular journal, research in a specialty or set of specialties—or should the unit of analysis be defined in some other way?

To what extent are the intellectual concerns of scientists working in the same field in fact related? The biological sciences are perhaps the most problematic in this regard. Recent developments in biology have split the biological sciences into a multiplicity of distinct fields; work done in one area may have little relation to or relevance for work done in other areas. Similar problems exist in many other fields: a solid-state physicist often has more intellectual interests in common with a physical chemist or a biophysicist than with a high-energy physicist; an experimental social psychologist in a sociology department more often than not has greater intellectual interests in common with some members of the psychology department than with the sociologists in her own department. Meaningful dissimilarities of interest may be as great within a single field as between fields.

The studies I will report have found similar degrees of consensus in fields such as physics and sociology, but this finding does not mean that consensus in the two fields has a similar source. For example, if an entire field is treated as the unit of analysis, we may find lower levels of agreement on some matters in physics than in sociology. But it is possible that physics may be a more highly specialized and differentiated field: high-energy physicists and solid-state physicists employ distinctly different theories and methods. Thus if both groups of physicists are considered part of the same unit for analytic purposes, the level of agreement may be reduced. Because sociology is a less clearly differentiated discipline than physics, researchers in many different areas of sociology may use the same theories and methods. The ideas of Goffman, Merton, or Weber and techniques such as regression analysis are used in many specialties. This uniform employment of diverse theoretical ideas and methods tends to create consensus across research areas, if only because sociologists in one research area can read, comprehend, and use work produced in most others. If, however, we were to take the research area as the unit of analysis, we might find different results. We might find more agreement within a research area such as high-energy physics than in any research area in sociology.

Although selecting the proper unit of analysis is crucial when doing research on consensus, it is unlikely that the results reported here are simply an artifact of the units of analysis. This is especially true of the data from the NSF peer review study, which came from specific programs, not from fields. Thus the data on chemistry were from the program of chemical dynamics, which is only one of seven subsections of the NSF chemistry section. However, the data from the social science, anthropology, included proposals submitted by both social anthropologists and physical anthropologists. The result for this study is therefore unlikely to have been an artifact of aggregation difficulties. I will return to this problem after presenting the results of my comparative research on consensus.

The Hierarchy of the Sciences?

Almost two hundred years ago Auguste Comte, the father of the history of science, set out what he called the "hierarchy of the sciences," maintaining that the sciences progress through ordained stages of development at quite different rates. For Comte, astronomy, the most general of all the sciences, develops first and is followed successively by physics, chemistry, biology, and finally sociology. The hierarchy of the sciences described not only the complexity of the phenomena studied by the different sciences but also their stage of intellectual development.[4]

Contemporary discussions of knowledge in the various sciences do not differ significantly from that of Comte. Two generations ago Conant (1950) proposed that the various branches of science differ in "degree of empiricism." Storer (1967) makes a distinction between the "hard" natural sciences and the "soft" social sciences. Kuhn ([1962] 1970) differentiates among the sciences by the extent to which they have a developed paradigm or shared theoretical structures and methodological approaches about which there is a high level of consensus. The sciences at the bottom of the hierarchy (for example, sociology) are assumed to exhibit less consensus and are frequently referred to as being in a "preparadigmatic" state.

Drawing on recent thought in this tradition, Zuckerman and Merton (1973a) proposed that the cognitive texture of scientific fields differs in the extent to which they are codified. Although they do not define the concept in detail, codification refers to the extent to which a field's paradigm or theoretical orientation is systematically developed. "Codification refers to the consolidation of empirical knowl-

edge into succinct and interdependent theoretical formulations"
(p. 507). Physics and biochemistry are hypothesized to be relatively
codified fields; geology and zoology are less so; sociology, anthropology, and political science are presumably among the least codified.
They further suggested that the degree of codification of knowledge
would be reflected in the level of agreement on the significance of
specific scientific contributions and on the importance of work of
particular scientists. Six variables underlie this concept of the hierarchy of the sciences (see Table 5.1).

Table 5.1 Characteristics of different types of science

Variable	Top of the hierarchy	Bottom of the hierarchy
1. Development of theory	Highly developed theory; research guided by a paradigm; high levels of codification	No theory or low-level generalization; preparadigmatic phase; low levels of codification
2. Quantification	Ideas expressed in mathematical language	Ideas expressed in words
3. Cognitive consensus	High levels of consensus on theory, methods, significance of problems, significance of an individual's contribution	Low levels of consensus on theory, methods, significance of problems, significance of an individual's contribution
4. Predictability	Ability to use theory to make verifiable predictions	Inability to make verifiable predictions
5. Rate of obsolescence	High proportion of references to recent work as older work becomes obsolete—indicating significant cumulation of knowledge	Low proportion of references to recent work as older work remains just as important as new work—indicating lack of significant cumulation of knowledge
6. Rate of growth	"Progress" or the rate at which new knowledge grows is relatively fast	"Progress" or the rate at which new knowledge grows is relatively slow

Source: S. Cole, 1983, p. 113, Table 1.

For the past two hundred years it has been assumed that there are differences among the sciences in levels of cognitive consensus. Highly codified fields such as physics are assumed to have substantially higher levels of agreement than are less codified fields such as sociology. Before presenting the empirical results of the studies designed to test this hypothesis, it is necessary to stress the conceptual distinction made in this book between knowledge at the research frontier and knowledge at the core. In their generalizations about knowledge in various sciences, historians, philosophers, and sociologists of science have tended to treat knowledge as if it were a uniform whole. There are clear-cut differences, however, in the types of knowledge found in fields such as physics and sociology. For example, in physics there are a small number of theories which are virtually universally accepted. The cornerstone of contemporary physics is the theory of quantum mechanics. Likewise, if we consider a specialty such as superconductivity research, almost all physicists working in this specialty recognize the 1957 theory of Bardeen, Cooper, and Shrieffer as the foundation of all contemporary work in the field. There seem to be no comparable examples in sociology.

Because obvious differences exist between the two fields, sociologists have assumed that we would find similar differences in all areas of knowledge. Because historians and philosophers of science naturally are interested in the most important knowledge, most of their theoretical formulations have been based upon a study of the core. This is true even of sophisticated contemporary historians and philosophers such as Kuhn and Laudan. Consider Laudan's (1984) comparison of levels of consensus in the natural and social sciences:

> To anyone working in the humanities or the social sciences, where debate and disagreement between rival factions are pandemic, the natural sciences present a tranquil scene indeed. For the most part, natural scientists working in any field or subfield tend to be in agreement about most of the assertions of their discipline. They will typically agree about many of the central phenomena to be explained and about the broad range of quantitative and experimental techniques appropriate for establishing "factual claims." Beyond this agreement about what is to be explained, there is usually agreement at the deeper level of explanatory and theoretical entities. Chemists, for instance, talk quite happily about atomic structure and subatomic particles. Geologists, at least for now, treat in a matter-of-fact fashion claims about the existence of massive subterranean plates whose motion is thought to produce most of the observable (i.e., surface) tectonic activity—claims that, three decades ago, would have been treated as hopelessly speculative. Biologists agree

about the general structure of DNA and about many of the general
mechanisms of evolution, even though few can be directly observed.
(p. 3)

Because our ideas about the structure of knowledge in various scientific fields have been heavily influenced by a reading of both historians
and philosophers of science, it is not surprising that most sociologists
of science have assumed that the level of consensus at the research
frontier of various fields will be correlated with the level of consensus
found at the core of the same fields.

Given the position of social constructivist sociologists of science that
nature plays no significant role in determining the scientific consensus, we would expect these sociologists to be less likely to accept the
standard assumptions about the ways in which the natural and the
social sciences differ. A large part of their work is aimed at showing
that the natural sciences do not correspond to our idealized images
based upon the traditional philosophy and history of science. Knorr-
Cetina (1981) has frequently argued that there are not substantial
differences between the natural and the social sciences. She maintains
that the methods utilized in the natural sciences are no more "objective" than those used in the social sciences: "The argument is not that
natural and technological scientists act like everyone else when they
talk to their peers or fight with their superiors in the organisational
hierarchy, but that their *methods* and procedures are sufficiently akin
to those of social science to cast doubt on the common distinction
between the two sciences" (p. 137).[5]

Mulkay (1978, p. 110) has also argued that the extent to which
there is a high level of cognitive consensus in the natural sciences has
been exaggerated in the accounts of writers such as Kuhn and Ziman.
But other sociologists of science who have been in general sympathetic to the approach of the constructivists continue to assume that
there are differences in cognitive consensus between the natural and
the social sciences. Brannigan (1981), for example, assumes that the
high levels of collaboration, higher acceptance rates in journals, and
"persistence of research groups and specialty networks" are indicators
of higher levels of consensus on theories and methods in the natural
sciences. Whitley (1984) assumes that there are higher levels of consensus in the natural sciences and offers explanations of why this is
so. One such explanation is that the social sciences deal with problems
of "everyday concern" and multiple audiences, thus introducing a
"plurality of standards and goals" (pp. 6, 26, 85).

Because most sociologists of science have tended to assume that there are substantial differences in the way in which knowledge is organized in the various fields, very little work has been done on this problem. The work most frequently cited to back up the assumption that there are consensus differences between the natural and the social sciences is that of Lodahl and Gordon (1972). Of particular interest was their survey of the *perceptions* among scientists in four fields (physics, chemistry, sociology, and political science) about the relative degree of paradigm development of seven scientific fields, including their own. After a brief explanatory introduction, they asked: "Please rank the following fields on the degree to which you feel there is consensus over paradigms (law, theory and methodology) within the field" (p. 59). The results conform exactly to the existing assumptions about the hierarchy of the sciences. Physics was perceived as having the highest level of paradigm development, followed by chemistry, biology, economics, psychology, political science, and sociology (with some disagreement about the order of the last two). Lodahl and Gordon reported no data on the *actual* levels of consensus in the various fields, nor did they ask about perceived levels of consensus on research frontier science. The study should be used simply as evidence that scientists in all fields strongly *believe* that a hierarchy exists and that the perceptions of the hierarchy among scientists in different fields are almost identical.

In a qualitative study of chemists and sociologists who were denied tenure in Ph.D.-granting departments, Rubin (1975) found chemists more likely to blame themselves and sociologists more likely to dispute the validity of the criteria used to evaluate them. This difference is attributed to assumed differences in the level of paradigm development and consensus in the two fields. But Rubin, too, assumed that there were differences in levels of consensus rather than actually measuring these levels.

In one of the most systematic studies of field differences, Hargens (1975) presents data on chemistry, mathematics, and political science. He makes a distinction between normative integration (cognitive consensus) and functional integration, or the degree to which scientists in a field are familiar with each other's work and interact with each other in a community. Using the data then available, he concluded that political science was less normatively integrated than the two other fields, but that mathematics was the least functionally integrated. But Hargens also had no direct measure of the level of consensus.

Journal Rejection Rates and Consensus

In a study of evaluation patterns in science, Zuckerman and Merton (1973b) observed significantly lower rejection rates for journals in the natural sciences than for those in the social sciences. This difference was taken as evidence of higher levels of consensus in the natural sciences, although Zuckerman and Merton point out that the results could be an artifact of noncognitive factors such as differing amounts of space available in the journals in the various fields. Hargens (1988a) extends the argument that the difference in rejection rates is evidence of differing levels of cognitive consensus in the various sciences. He presents data similar to that collected by Zuckerman and Merton and argues that the high rejection rates of social science journals results from the lack of cognitive consensus among contributors and reviewers. He claims that the low rejection rates in fields such as physics result from a high consensus between contributors and reviewers.

Although the argument that differing rejection rates result from differing levels of consensus is appealing because it supports the prior assumption of most sociologists about the cognitive differences between the hard and soft sciences, I do not find it convincing.[6] Journal rejection rates may be influenced by consensus, but they are also likely to be influenced by other variables. The first variable which may influence rejection rates is the amount of space available in journals. Hargens (1988a) attempts to prove that this variable is not significant by showing that changes in the number of submissions over time does not explain much variance in the acceptance rate over time. But he has no data on changes in the number of pages published in each journal. If increases in submissions were accompanied by increases in the number of pages published, it would not be surprising that the number of submissions variable had little influence on the rejection rate. Even if there were no change in the number of pages published, it would not be surprising to find no correlation between change in submissions and change in rejection rate *independent* of the rejection rate at Time 1, because the variables which caused the differences in rejection rates, including differences in space shortage, were already present at Time 1. In his 1975 monograph Hargens presented direct evidence that showed that the ratio of mean circulation to mean number of articles published was more than twice as high in social science journals than in chemistry journals, and that backlogs of unpublished articles were substantially greater in social science journals.

Beyer (1978) presents data which leads her to conclude that the higher rejection rate of social science journals is at least in part a result of scarcity of space. For her samples of journals in physics, chemistry, sociology, and political science she computes ratios of the average journal circulation divided by the number of articles published. These were 7.6 for physics, 9.8 for chemistry, 114.8 for sociology, and 195.7 for political science (p. 79). And, Beyer recognizes, her data substantially underestimate the disparity in space availability. This is because she counted only the "A series" of journals, such as *Physical Review (PR)*, which publish additional series for specialties such as solid-state physics. In fact Beyer's seven physics journals published an average of 4,000 pages in 1973. Meyer (1979) analyzes data on all the series published by both *PR* and *Physical Review Letters (PRL)*. In 1971 these two journals published a total of 25,537 pages and in 1977 a total of 32,064.[7] In 1977 the *American Sociological Review (ASR)* and the *American Journal of Sociology (AJS)* published a total of 1,997 pages devoted to articles.[8]

Beyer asked the editors of the journals whether they favored the expansion in size of their journal and whether they would be able to maintain the quality of published articles if such an expansion took place. She found that most of the social science editors, but only a minority of natural science editors, favored expanding the size of their journals. Most of the social science editors did not think that expanding the size of their journal would reduce its quality. Beyer (1978) concludes:

> We were interested in the question of whether publication was delayed in the social sciences because of an overload of publishable manuscripts that could not be accommodated in available space. There are a variety of ways to deal with overload in an information system. One way, of course, is by omission: editors can simply reject more articles. *This method is used by the social science journals.* While some purists will argue that the quality of social science research is low and this is why rejection rates are so high, data already reported suggest that many of these social science editors feel they could expand the size of their journals and maintain the same quality standards as they have at present. This strongly suggests that the present criteria between acceptance and rejection must be very narrow for these journals. (p. 81, italics added)

These data suggest that the amount of space available cannot be discounted as an important reason for the higher rejection rates in the social sciences.

Those who argue that journal rejection rates result from the level of disciplinary consensus often make the implicit assumption that in the social sciences most articles submitted to the journals are of poor quality and not "really" publishable, owing to lack of agreed-upon criteria. There are two problems with this assumption. First, it assumes that the acceptance of most of the articles submitted to the natural science journals indicates that these articles deserve to be published. Many studies of citation patterns, however, have shown that the bulk of articles published in physics journals, for example, are rarely if ever cited (Meyer, 1979). There is also qualitative evidence that natural scientists are just as likely to disparage the quality of articles in their journals as are social scientists. Mulkay and Williams (1971), in their study of physicists in England, report that "*all* our respondents thought that the vast majority of papers in the journals which they read were of poor quality or of little significance" (p. 74). Second is the assumption that the articles rejected by the social science journals deserve to be rejected. Stinchcombe and Ofshe (1969) conduct an interesting analysis in which they assume that the ability to evaluate articles submitted for publication is about the same as other measurement processes based upon qualitative materials. The validity of a judgment of an article would therefore be about .70. They then show that, given this validity coefficient and the decision to publish only about 15 percent of submitted articles, almost as many papers which "truly" deserve to be published will be rejected as accepted. This finding points to the crucial nature of the decision prior to evaluation to accept only a fixed proportion of articles submitted. Later research (Cicchetti, 1991) suggests that the validity of judgment is considerably below .70, which would mean that substantially less than half of the articles which "truly" deserve to be published will be accepted.

A second variable which could influence journal rejection rates is field-specific norms concerning publication. Hargens (1988a) is aware of this variable, pointing out that "Zuckerman and Merton [1973b] note that journals with high acceptance rates tend to use procedures that *presuppose* that submitted papers should be published and to minimize the chance that worthy papers will be rejected" (p. 141, italics added; see also Bakanic, McPhail, and Simon, 1987, p. 633). Some fields, such as physics, have a norm that submitted articles should be published unless they are wrong. Ziman (1968) has commented on this norm: "The general procedure is to allow all work that is *appar-*

ently valid to be published: time and further research will eventually separate the true from the false. A mistaken observation will be repeated, and the discrepancy noted and corrected; a bad piece of logic or of calculation will be reworded and put right by some other person in due course" (p. 55, italics added). Physics journals prefer to make "Type I" errors of accepting unimportant work rather than "Type II" errors of rejecting potentially important work. This policy often leads to the publication of trivial articles with little or no theoretical significance, deficits which are frequently cited by referees in social science fields in rejecting articles. Other fields, such as sociology in the United States, follow a norm of rejecting an article unless it represents a significant contribution to knowledge. Sociologists prefer to make Type II errors.

Hargens (1988a) concludes that "editors [of social science journals] rarely report that they have been forced to reject more papers than the reviewing process dictates" (p. 141). He offers no qualitative or quantitative evidence to support this statement, which is contradicted by a letter recently sent by the editor of one of the most prestigious sociology journals to a contributor. The editor had sent the paper to two reviewers, both of whom recommended publication and expressed highly favorable opinions of the paper. Along with the verbatim reviews, the editor sent the contributor a letter of rejection containing the following sentence: "I have enormous confidence in the two reviewers who gave you positive reviews, but [name of journal] has limited space and *I continually confront the dilemma of selecting among good papers*" (italics added). It is impossible to imagine a contributor to *PR* or *PRL* receiving such a response. But it is not unusual for contributors to *Science*, a journal with an extremely high rejection rate, to receive such a letter.

In the case of most articles submitted to journals in sociology and the other social sciences with high rejection rates, the editor will have received negative reports from referees which back up the rejection. But these negative reviews may be strongly influenced by the reviewers' knowledge of the norm that they are *supposed* to reject most papers they are sent. Thus even if editors should report rarely having to reject deserving papers, this information should not be surprising if we assume that the reviewers, aware of and influenced by the norms, provide the editors with the opportunity to conclude that the articles submitted are not publishable. Some editors explicitly inform the reviewers of the rejection norm. The editor of *Social Forces (SF)*, one of the most prestigious journals in American sociology, sends all reviewers a letter in which the second paragraph is as follows: "In making your recommendation to the Editor, please keep in mind that we have space for only about 15% of the manuscripts submitted." A reviewer receiving such a letter may look for reason to reject.

Qualitative data support the publication-norm hypothesis. For example, one of the most important sociology journals in Poland, *Studia Sociologiczne,* accepts a majority of the papers submitted.[9] In Poland, sociologists do not subscribe to the norm that articles should be rejected unless they are a significant contribution. Hargens would have to interpret these data to mean that there is more cognitive consensus among Polish sociologists than among American sociologists.

A third variable not considered by Hargens that might influence rejection rates is the diffuseness of a field's journal system. In physics, the two leading journals publish a substantial portion of all literature in the field; in sociology, the two leading journals publish only a small fraction of all literature. Sociologists must therefore submit their articles to one journal after another until the article is published or the author gives up. Bakanic, McPhail, and Simon (1987) point out that of all manuscripts submitted to *ASR* nearly 29 percent were eventually published in that journal. Garvey, Lin, and Nelson (1970) note that more than 20 percent of articles published in "core" social science journals have previously been rejected by one or more journals. Thus some articles rejected by *ASR* will be published by the *AJS, SF,* or some other core journal. It is probable that at least half the articles submitted to *ASR* are eventually published in a core journal and that three quarters of them are eventually published somewhere. If *ASR, AJS, SF,* and all the other journals published by the American Sociological Association were one journal, a situation similar to that in physics, it is probable that the rejection rate in sociology would not differ significantly from that in some of the natural sciences.

Hargens (1988a) treats these variables by arguing that they are intervening between consensus and rejection rates, that is, that they are caused by differences in consensus. For example, that *ASR* and *AJS* publish only a small portion of the literature in sociology and that *PR* and *PRL* publish a large portion of the literature in physics is, he implies, a result of differences in consensus between the two fields. Until there is some direct measure of consensus, however, this argument simply restates the assumption that the natural sciences have more consensus than do the social sciences.

We may conclude that most sociological studies dealing with field differences in consensus either have assumed the differences to exist, frequently using Lodahl and Gordon (1972) for authority, or have used indirect data such as journal rejection rates.

Consensus and the Operation of the Reward System

Here I shall report the results of a series of studies I conducted aimed specifically at examining the extent to which cognitive consensus exists in the various scientific fields. Previous research on physicists (J. R. Cole and S. Cole, 1973), for example, led to the conclusion that quality of published research was the most important determinant of three forms of recognition: receipt of honorific awards, appointment to professorships at prestigious departments, and having one's work widely known. I also found that the quality of newly published papers in physics explained practically all the variance on their reception (S. Cole, 1970). Papers of high quality were favorably received regardless of the physicist's location in the stratification system or any other personal or social characteristic. We (J. R. Cole and S. Cole, 1973) concluded that physicists who publish high-quality work will probably be rewarded no matter who they are or where they are located.[10] A study was designed to test the validity of this conclusion for an additional sample of physicists and for scientists working in four other fields: chemistry, biochemistry, psychology, and sociology (S. Cole, 1978).

I hypothesized that the extent to which a field is characterized by consensus should have a significant influence on the structure of its reward system. In high-consensus fields such as physics and chemistry there would be substantial agreement on who is doing important work, and the sheer quantity of publications would therefore have an insignificant effect on the distribution of rewards. In low-consensus fields such as psychology and sociology there should be less agreement on who is doing important work. In the absence of such agreement, the sheer quantity of publications would have a greater influence on the distribution of rewards. It may be relatively difficult to reach consensus on whether or not a paper is a significant or insignificant contribution, but it is relatively easy to count the number of papers. In this study I was not measuring consensus directly. Rather I reasoned that if differences in cognitive consensus exist they should have a specified observable effect on the social organization of the disciplines.

The data for this study were based on samples of scientists who were promoted from associate to full professor at Ph.D.-granting institutions between 1964 and 1969 in the following fields: biochemistry, chemistry, physics, psychology, and sociology.[11] The study had three dependent variables: the perceived quality of a scientist's work and the scientist's visibility, both measured by questionnaires, and the rank of the scientist's academic department as determined by Roose and Anderson (1971). The independent variables were the rank of the scientist's doctoral department and the quality and quantity of the scientist's work. Quality of work was measured by citations and quantity by publication counts.

The hypothesis was that the quantity of work would have a greater impact than quality of work on rewards in the less codified social sciences than in the more codified natural sciences. In all five fields, there were moderate to strong correlations between the quantity of published research and its quality, ranging from .53 for psychology to .72 for chemistry. Although quality and quantity of published research were correlated in all fields, the correlation was not perfect, which indicated that in each field there were some scientists who produced many relatively insignificant papers and others who produced only a few highly significant papers.[12] But how does quantity and quality of published research influence the distribution of rewards in each field? In each of the five fields and in every comparison made, the correlation between quality of output and recognition was at least slightly higher than that between quantity and recognition. For example, in sociology the correlation between quality and rank of department was .51 and that between quantity and rank of department .42. What is more important is that there were no systematic differences between the social and the natural sciences. On the basis of these data, at least, it could not be said that quantity is a more important determinant of success in the social sciences than it is in the natural sciences.

Several different techniques were employed to see if there were any significant differences in the ways in which the reward systems in the five disciplines operated. A multiple regression analysis of the determinants of visibility was performed, using four independent variables: quality of publications, quantity of publications, perceived quality of work (measured by questionnaire data), and rank of the scientist's academic department. This analysis showed not only no differences between the social and the natural sciences but virtually no differences among any of the five fields. The multiple R achieved for the five fields varied from a low of .74 for biochemistry to a high of .78 for chemistry. Apparently the determinants of visibility are the same in these five fields.

As a final test of the extent to which there might be differences in the reward systems of the disciplines I developed a causal model depicting the operation of the reward system. Elsewhere (S. Cole, 1978) I have analyzed the substantive details that this model reveals about the operation of scientific reward systems. Of interest here is the extent to which the reward systems of the less codified social sciences might differ from that of the more codified natural sciences. To what extent can we assume that the five fields are samples drawn from the same population? The analysis reported in S. Cole (1978) led to the conclusion that there are no significant differences in the ways in which the reward system in these five fields operate.[13]

At least two conclusions may be drawn from this study. First, it is possible that although substantial differences in consensus do exist in the various fields, these differences may not have the expected influence on the reward system. In this case the failure to find the quantity of work a more significant influence on rewards in the social sciences than in the natural sciences would not lead to the conclusion that the levels of cognitive consensus are similar. A second conclusion is that the levels of cognitive consensus in the various fields, in so far as work at the research frontier is concerned, are similar. This study, however, as are any of the studies in isolation, is incapable of fully answering this question. It is only by looking at all of the empirical studies that we can approach a conclusion on this issue.

Age and Scientific Performance

A strong component of the mythology of science is that science is a young person's game. It is commonly believed that scientific creativity declines with age. In a series of empirical investigations aimed at testing this hypothesis I found only a slight curvilinear relationship between the quality and quantity of publications and age.[14] This relationship is molded by the structure of the scientific reward system (S. Cole, 1979). Here I will focus on one aspect of the investigations of the relationship between age and scientific performance: the extent to which this relationship varies with the degree of codification of the discipline.

In their paper on age in science, Zuckerman and Merton (1973a) hypothesize that in a more codified field, such as physics, it should be easier for young scientists to make significant contributions than in a less codified field such as sociology, for two reasons. First is the method of gaining competence in fields of varying degrees of codification. Because in the highly codified fields knowledge is compacted, graduate students can quickly learn the current state of their field from textbooks and begin work on the research frontier while still in school (Kuhn, [1962] 1970). In the less codified fields knowledge is not as compacted, and far greater experience is needed to gain competence. "In these [the less codified fields], scientists must get command of a mass of descriptive facts and of low-level theories whose implications are not well understood" (Zuckerman and Merton, 1973a, p. 507). Zuckerman and Merton conclude that "codification facilitates mastery of a field by linking basic ideas in a theoretical framework and by reducing the volume of factual information that is required in order to do significant research. This should lead scientists in the more codified fields to qualify earlier for work at the research front" (p. 510).

The second reason it is said to be easier for young scientists to make significant contributions in the highly codified fields involves variation in the social process through which contributions are identified as significant. In the highly codified sciences there is presumably greater consensus, and it may therefore be easier to identify new contributions as important or unimportant. In the less codified fields the identification of a contribution as significant is more dependent upon the reputation of its author. "In these less codified disciplines, the personal and social attributes of scientists are more likely to influence the visibility of their ideas and the reception accorded them. As a result, work by younger scientists who, on the average, are less widely known in the field, will have less chance of being noticed in the less codified sciences" (Zuckerman and Merton, 1973a, p. 516). For these two reasons, ease in moving quickly to the research front and ease in having an important contribution identified as such, younger scientists should be more likely to make important contributions in the natural sciences than their age peers in the social sciences.

In order to examine the codification hypothesis I collected data on random samples of scientists employed in Ph.D.-granting departments in six different fields.[15] Of the six fields included in the age studies, physics and mathematics are presumably the most highly codified, followed by chemistry, geology, psychology, and sociology. If the codification hypothesis is correct, we should find a higher proportion of young scientists making important contributions in physics, mathematics, and chemistry than in the other fields. Because the distribution of citations to scientists working in the various fields differs greatly, in order to compare scientists in different fields I had to standardize the data separately for each field. All scientists with standard scores greater than zero could be classified as having made *relatively* high quality contributions to their fields. The proportions of young scientists (those under the age of 35) whose work published between 1965 and 1969 received a relatively high number of citations in 1971 (a positive standard score) were as follows: chemistry (29%), geology (38%), mathematics (11%), physics (25%), psychology (24%), and sociology (13%). If we examine these proportions, we find little support for the codification hypothesis. Geology is the field in which the highest proportion of young scientists has made significant contributions, and mathematics is the field in which the lowest proportion of young scientists has made significant contributions. Furthermore, young psychologists are almost as likely to make important contributions as are young physicists and chemists.

Let us consider some possible flaws in this empirical test. It might be argued that the codification hypothesis is meant to apply only to contributions of high import and that in the more codified fields it will be easier for younger scientists to make truly outstanding contributions. However, no matter how high the cutoff point was set for classification of important contributions, no systematic differences among fields were found. I decided, however, to perform an additional test of the codification hypothesis and examine the ages at which the most eminent scientists in various fields make their first significant contribution.

Two sets of data were used. The first comes from the same questionnaires used in the comparative reward system study already discussed.[16] Respondents were asked to name the five scientists who they felt had contributed most to their field in the past two decades. I examined the ages at which the ten most frequently mentioned scientists in each field made their first significant contribution. In this procedure it is important that the same criteria for defining an important contribution be used in each field. The first approximation to a more definitive test was to record the age at which each scientist published his or her first paper which received five or more citations in the 1971 *Science Citation Index (SCI)*. For example, the earliest paper by C. N. Yang, the Nobel laureate in physics, which received five or more citations in 1971 was published in 1951, when Yang was 29.

The data collected offer slightly more support for the codification hypothesis (see Table 5.2). Physicists do make their first important contributions at a slightly younger age than do other scientists. In the relatively highly codified field of biochemistry, however, scientists take just as long on the average to make their first important contribution as in sociology. And in experimental psychology, which is presumably more codified than clinical psychology, scientists are no younger when they make their first significant contribution.

I replicated this test on another set of data: lists of the most frequently cited physicists in the 1965 volume of the *Physical Review* and the most frequently cited sociologists in the 1970 volumes of the *ASR* and *SF* (Oromaner, 1972).[17] These lists can be considered to be samples of the scientists currently producing the most useful work in each of the fields. At what age did each scientist make his or her first significant contribution? Using the cutoff point of five citations I found that the mean biological age at which physicists made their first important contribution was 31, as compared with a mean age of 34 for the sociologists (see Table 5.3).

It is possible, however, that the results might be influenced by differences in the age at which physicists and sociologists earn their doctorates. The mean age at which physicists earn their doctorates is 26, as compared with 28 for sociologists. Although it might be claimed that this merely reflects the shorter time required to learn the more codified field, it also might be a result of the different requirements for the doctorate. In physics, doctoral dissertations generally concern

Table 5.2 Mean age at first publication receiving 5 or more citations for five fields

Field	Mean age	N^a
Biochemistry	35	11
Chemistry	30	10
Physics	27	10
Psychology (clinical)	34	10
Psychology (experimental)	34	11
Sociology	34	10

Source: Adapted from S. Cole, 1979, p. 974, Table 9.

a. In biochemistry and experimental psychology two scientists were tied for tenth place. Both were included—giving 11 scientists for these two fields.

Table 5.3 Mean age at first publication receiving 5 or more citations for most-cited physicists and sociologists

	Physics	N	Sociology	N
Mean biological age	31	(50)	34	(31)
Mean professional age	5.0	(50)	6.0	(31)
Mean professional age (scientists under 60 in 1971)	4.9	(48)	4.8	(24)
Mean professional age (scientists under 50 in 1971)	3.9	(42)	3.4	(17)

Source: S. Cole, 1979, p. 975, Table 10.

shorter, more discrete problems than do dissertations in sociology, which are usually more equivalent to books than to articles. At any rate, most scientists in both fields generally have little opportunity to publish prior to receipt of their doctorate. Because physicists are younger at receipt of the doctorate, to make the two fields more comparable we should examine the professional age (years since earning the doctorate) at which scientists in the two fields make their first important contribution.[18] It took the physicists on the average 5.0 years from receipt of the doctorate and the sociologists 6.0 years to make their first significant contribution. Because citations made in 1971 may not be a very reliable indicator of the quality of very old work, we can omit from the sample all scientists aged 60 or older in 1971. The mean professional age at which the physicists under 60 make their first important contribution is 4.9, as compared with 4.8 for the sociologists. When scientists 50 or older in 1971 are excluded we get a mean professional age of 3.9 for physicists and 3.4 for sociologists. No matter which comparisons we make, we must at least tentatively conclude that there is no meaningful difference in the age at which scientists in a highly codified field such as physics and a relatively uncodified one such as sociology make their first significant contribution.

This study, like the comparative reward system study, raises questions about the hierarchy hypothesis. If indeed there are differences in the level of codification and cognitive consensus at the research frontiers of the various fields we studied, then the hypothesis of Zuckerman and Merton that such differences affect the relationship between age and scientific productivity would seem to be incorrect. Or perhaps, when knowledge at the research frontier is considered, the hypothesized consensus differences among fields do not exist.

Consensus on Evaluating Scientists

Other studies were designed to examine the level of consensus more directly (S. Cole, 1978).[19] The first set of data I analyzed was also drawn from the questionnaires used in the comparative reward system study. The questionnaire for each of the five fields listed the names of sixty scientists, each recently promoted to full professor at an American Ph.D.-granting institution. For each name, the raters were asked to respond to the following: "Please indicate the relative importance of the work of the following scientist: has made very important contributions, has made above average contributions, has made average contributions, work has been relatively unimportant, unfamiliar with work but have heard of this scientist, have never heard of this scientist."[20]

As a measure of consensus I computed the standard deviation of the ratings received by each of the 60 scientists. The smaller this standard deviation, the greater the agreement in evaluation. I then computed the mean standard deviation for each of the five fields. Differences among fields followed the expected pattern, with physics displaying the most agreement on evaluation and sociology the least (see Table 5.4). The differences, however, were quite small and only that between physics and sociology was statistically significant at the .05 level. The difference between any other pair of fields, for example, sociology and chemistry, was statistically insignificant. The results of this preliminary investigation suggest the hypothesis that there may not be major differences among the fields in the degree of consensus in evaluating the contributions of samples of scientists.

Other data enabled me to measure this form of consensus in a

Table 5.4 Consensus on evaluating scientists by field (60 scientists in each field)

Field	Mean standard deviation of ratings	N raters
Biochemistry	.71	107
Chemistry	.69	111
Physics	.63	96
Psychology	.74	182
Sociology	.76	145

Source: S. Cole, 1983, p. 120, Table 3.

second way. On the same questionnaires, I asked the raters to list the five scientists who had contributed most to their discipline in the past two decades. Presumably, scientists in highly codified fields would show greater agreement in this choice than those in less codified fields. I tabulated the results in three different ways: the percentage of total mentions going to the five most frequently named scientists, the percentage going to the ten most frequently mentioned scientists, and the number of different scientists mentioned divided by the total number of mentions. The higher the first two proportions and the lower the last, the greater the approximation to consensus. The results, reported in Table 5.5, again suggest that there may be no systematic differences in this type of consensus between fields of varying levels of codification. In fact, if we use the percentage of mentions going to discrete names as a measure, sociology has the highest degree of agreement, and chemistry the lowest.

Some data on the extent to which reviewers of grant proposals agreed on an assessment of the scientific reputations of the principal investigators were presented in Chapter 4 (see Table 4.7). Recall that reviewers of proposals in chemical dynamics, economics, and solid-state physics were asked to evaluate how much the principal investigator had contributed to his or her "field" and "specialty." The data revealed no more agreement among reviewers on reputations than on the content of the proposals. On the question of whether there was greater agreement on reputations among the natural scientists (chemical dynamics and solid-state physics) than among the social scientists (economists), the data indicate no significant differences in level of consensus among the three programs studied.

Table 5.5 Consensus on which scientists have made the most important recent contributions, by field (60 scientists in each field)

| Field | Percentage of mentions received by | | | |
	Five most mentioned names	Ten most mentioned names	Discrete names	*N* mentions
Biochemistry	41	56	21	473
Chemistry	34	47	39	377
Physics	47	65	23	375
Psychology	32	44	36	435
Sociology	36	52	19	661

Source: S. Cole, 1983, p. 120, Table 4.

Another, more concrete, measure of consensus is the distribution of citations in a scientific journal. In fields characterized by a high level of intellectual agreement, we would expect to find a heavy concentration of references to a relatively small number of papers and authors. The distribution of references in fields characterized by lack of agreement should more closely approximate a random distribution. For example, in a field such as physics, where there is thought to be a heavy concentration of research effort on a few important problems, we should expect to see a higher concentration of citations than in a field such as sociology, where the range of problems studied is much wider. Using the Gini coefficient as a measure of concentration (the higher the coefficient, the greater the concentration), my colleagues and I analyzed the distribution of citations in 1971 in 108 leading scientific journals.[21]

We found significant differences in the concentration of citations in the various journals (see Table 5.6). But whether the article or the author is used as the unit of analysis, relatively small differences are found in the mean Gini coefficient for the various fields. When the article is the unit of analysis, the two social sciences, psychology and sociology, have about the same mean Gini coefficients as geology and mathematics. Examination of the complete range of data shows that, for example, the psychology journal with the highest Gini coefficient, *Journal of the Experimental Analysis of Behavior,* had a higher coefficient than any journal in chemistry, geology, or mathematics. Considerable differences in the coefficients occur for different journals in the same field. In chemistry, for instance, with the article as the unit of analysis,

Table 5.6 Concentrations of citations, by article and by author, to research articles in selected fields

Field	Gini coefficients by cited article		Gini coefficients by cited author		N journals
	Mean	Range	Mean	Range	
Biochemistry	.21	.05–.34	.44	.32–.51	10
Chemistry	.15	.06–.27	.46	.30–.56	12
Geology	.10	.04–.23	.40	.29–.53	7
Mathematics	.09	.06–.13	.38	.33–.43	6
Physics	.18	.06–.35	.48	.37–.62	10
Psychology	.16	.05–.29	.42	.32–.54	8
Sociology	.09	.05–.11	.34	.30–.39	7

Source: Adapted from S. Cole, 1983, p. 126, Table 8.
Note: Specialties and fields with fewer than five journals available were omitted.

Analytical Chemistry has a Gini coefficient of .06, whereas the *Journal of Chemical Physics* has a coefficient of .27. When the author is used as the unit of analysis, the Gini coefficients for chemistry journals vary from a low of .30 to a high of .56. Similar variances are found in other fields that we intuitively believe to have a high level of codification; significant differences in coefficients are also obtained for journals in fields that we believe to have a lower level of codification. A good example is psychology, where the scores vary between .05 and .29 with the article used as the unit of analysis, and between .32 and .54 with the author as the unit. When the two social sciences were taken as one group and all other fields as another, an analysis of variance showed that the differences were not statistically significant with the article as the unit of analysis. When the author was used instead, the difference between the two groups was significant at the .01 level, but the variable "field" explained only 10 percent of the total variance on the Gini coefficients.[22]

Some evidence exists to support the hypothesis that consensus of the type measured by a Gini coefficient is related to codification. A field such as sociology generally has lower scores than a field such as physics. Specialties within a field that we intuitively believe to have lower levels of codification also generally have lower Gini coefficients. Thus in psychology, a clinical journal has a lower score than an experimental journal. In physics, optics and acoustics have lower scores than other specialties, and within the sections of the *Physical Review* itself, particle physics has a higher score than the other major specialties. In chemistry, analytic chemistry, which we intuitively believe to be less codified than other specialties, does indeed have a relatively low Gini coefficient. The *Journal of Geophysical Research*, which might just as easily be classified as a physics journal, has a considerably higher coefficient than any of the other geology journals.

The Gini coefficients of citation concentrations in scientific journals suggest that variations in agreement there may be due less to codification than to the scope of the research published in the various journals. The more specialized and restricted in subject matter or orientation the journal is, the higher the Gini coefficient. Using the article as the unit of analysis, it is instructive to compare, for example, the coefficients for the *Physical Review Letters* and for the four sections of *Physical Review*. These journals publish essentially the same research, done by the same scientists. But the *Letters*, which covers research from all specialties, shows a lower level of agreement (Gini coefficient = .16) than the sections of the *Review,* which divides up research by broad areas within the field. The Gini coefficients for the different sections of the *Physical Review* range from .35 to .21. The field of psychology provides another example: the .54 coefficient obtained for the *Journal of the Experimental Analysis of Behavior* is one of the highest among the 108 journals. This journal publishes only research done in a limited area of psychology.

Consensus among Peer Reviewers

The last set of data comparing cognitive consensus in the various scientific fields is drawn from the study of peer review at the National Science Foundation described in Chapter 4. Here I shall use the data collected in both phases of the research to compare the level of consensus among social science and natural science reviewers of proposals submitted to NSF. Recall that in Phase I we collected data on applications made to the NSF in fiscal year 1975 for ten different scientific programs: algebra, anthropology, biochemistry, chemical dynamics, ecology, economics, fluid mechanics, geophysics, meteorology, and solid-state physics. For each of these programs we had information on 100 grant applications, about half of which were funded. A grant proposal gives very concrete plans for research. Evaluating the proposal involves deciding whether the problem outlined is a significant one, determining whether the specific plans to investigate it are feasible, and evaluating the competence of the principal investigator. If there is indeed a higher level of cognitive consensus in the more codified fields at the top of the hierarchy than in those at the bottom of the hierarchy, we would expect to see greater consensus among reviewers of research proposals in the highly codified than in the less codified fields.

For the Phase I data I used the standard deviation of the ratings for each proposal as a measure of consensus (see Table 5.7). I then computed the mean standard deviation for the 100 proposals in each field. This mean ranged from a low of .31 in algebra to a high of .69 in ecology and meteorology. The results obtained from this analysis support the conclusions suggested by the analysis of variance performed in Chapter 4. Once again economics proved to be the field with the lowest level of disagreement.

Phase II of the peer review study was extensively discussed in Chapter 4, but we may use those data here to replicate the results on

Table 5.7 Consensus among reviewers of scientific proposals submitted to the NSF

Program	Mean standard deviation of reviewers' ratings	Mean of reviewers' ratings	Coefficient of variation
Algebra	.31	2.1	.15
Anthropology	.59	2.5	.24
Biochemistry	.60	2.6	.23
Chemical dynamics	.42	2.3	.18
Ecology	.69	2.3	.30
Economics	.34	2.6	.13
Fluid mechanics	.61	2.8	.22
Geophysics	.61	2.4	.25
Meteorology	.69	2.8	.25
Solid-state physics	.35	2.2	.16

Source: Adapted from S. Cole, 1983, p. 123, Table 6.
Note: Data for each program are based on 100 proposals.

consensus from Phase I. For each field, chemical dynamics, economics, and solid-state physics, we had three sets of reviewers: the original NSF reviewers, the COSPUP nonblinded reviewers, and the COSPUP blinded reviewers. If we consider the standard deviation of the reviewer variances for each of these groups of raters (see Table 5.8), the Phase II data suggest that there was a slightly higher level of consensus among the reviewers in the two natural sciences than there was in economics. Although the differences are statistically significant, they are small in size when we consider that the difference between a "very good" rating and a "good" rating is 10 points. The difference in the standard deviations between economics and the two natural sciences was about 2.5 points. In interpreting these data we must also keep in mind that the chemical dynamics and solid-state

Table 5.8 Standard deviation of reviewer ratings in three programs

Program	NSF	COSPUP nonblinded	COSPUP blinded
Chemical dynamics	7.5	7.5	9.1
Economics	9.4	9.8	10.1
Solid-state physics	7.0	7.1	8.5

physics programs included only work in a specialty; the economics program included work in an entire field.

In what is perhaps the most comprehensive analysis of data on peer review reliability, Cicchetti (1991) presents data from a secondary analysis of the Phase II experiment. After an extensive discussion of the statistics most appropriate for measuring consensus, Cicchetti selects the intraclass correlation coefficient (R_i, Model III; see Cicchetti, 1991, pp. 134–135). Using this measure he finds the level of consensus in economics slightly higher than that in the two natural sciences in all three groups of reviewers (i.e., NSF reviewers, COSPUP nonblind reviewers, and COSPUP blind reviewers). And when he combines the three sets of reviewers he finds R_i, to be .16 for chemical dynamics, .34 for solid-state physics, and .44 for economics. In general, both my analysis of the peer review data and the secondary analysis by Cicchetti lead to the conclusion that there are not large differences in levels of agreement on research proposals between fields located at the top of the hierarchy and those located in a lower position in the hierarchy.[23]

Core Knowledge

The data reported in the preceding four sections were drawn from studies which attempted to measure consensus at the research frontiers of the various scientific fields. Up to this point I have assumed that the differences hypothesized to exist between the natural and the social sciences by Kuhn, Zuckerman and Merton, and many others do apply to the core of these disciplines.[24] In an attempt to provide some data on this topic, I have conducted a preliminary investigation of undergraduate textbooks in chemistry, physics, and sociology. If we assume that textbooks represent the content of the core of a discipline, we may examine references in a textbook in order to describe the nature of that core.

In this preliminary study I examined only two variables: the age distribution of references, and the total number of references. The differences between the chemistry and physics textbooks, on the one hand, and the sociology textbooks, on the other, are striking (see Table 5.9). The overwhelming preponderance of material discussed in undergraduate chemistry and physics textbooks was produced prior to 1960. Only 6 percent of the references in chemistry texts and 3 percent of those in physics texts were to work published after

Table 5.9 Distribution by age of references in undergraduate texts
(in percent)

	Chemistry	Physics	Sociology
Before 1700	7	16	
1700–1799	10	8	
1800–1849	12	16	
1850–1899	26	17	1
1900–1919	12	10	1
1920–1939	19	19	4
1940–1959	8	11	19
Post-1959	6	3	75
Total	100	100	100
Mean *N* references			
per text	100	74	800

Source: S. Cole, 1983, p. 133, Table 10. Chemistry data based on W. L. Masterton and E. J. Slowinski, *Chemical Principles* (1977); and R. E. Dickenson et al., *Chemical Principles,* 3d ed. (1979). Physics data based on G. Shortley and D. Williams, *Elements of Physics,* 3d ed. (1961); W. Thumm and D. B. Tilley, *Physics: A Modern Approach* (1970); M. Alonso and E. J. Fenn, *Physics* (1970); K. Greider, *Invitation to Physics* (1973); and F. W. Sears, M. W. Zernansky, and H. D. Young, *University Physics,* 5th ed. (1976). Sociology data based on N. Goodman and G. Marx, *Society Today,* 3d ed. (1978); and I. Robertson, *Sociology* (1978).

1959. The exact opposite is true of sociology textbooks. Fully 75 percent of all references were to work published after 1959. Although it might be argued that chemistry and physics are indeed older disciplines than sociology, it is clear that any difference in the distribution of the age of literature in these fields could not possibly account for this dramatic difference in the age of references in textbooks. The great majority of all literature in chemistry and physics has been produced in the post–World War II period, owing to the very rapid exponential growth of these sciences since 1945 (Price, 1963). We should also note the large difference between the chemistry and physics texts, on the one hand, and the sociology texts, on the other, in the total number of references cited. Whereas authors of natural science textbooks generally cite approximately 100 articles or books in their texts, authors of sociology textbooks cite an average of 800 works.

Although data must be collected for a wider variety of fields and also for graduate textbooks in those fields where they exist, the data in Table 5.9 allow us to speculate about the differences between the core of knowledge in fields such as physics and chemistry and the core in fields such as sociology. As indicated by what is taught to undergraduate students, the core in fields such as chemistry and physics is made up of a small number of theories and exemplars which have been produced by scientists over the course of several hundred years. There is almost universal agreement about which are the most important theories and exemplars. The list of references in the five physics textbooks examined were characterized by substantial overlap, with a majority of names appearing in all five texts. The same was true for the two chemistry texts examined. In the sociology texts, most references were to a wide variety of empirical studies on topics of current interest. Practically all of the studies had been conducted since 1960. The textbooks in the two natural sciences pay little attention to work at the research frontier; those in sociology pay primary attention to such work, probably because in sociology the core of knowledge is very small and the frontier is relatively large.

A qualitative comparison of textbooks published in the 1970s in physics and chemistry with ones published in the 1950s indicated that the material covered in the 1950s and in the 1970s is essentially the same. This is not true of sociology. Sociology textbooks produced in the early 1950s relied heavily on recently published work, but textbooks published in the 1970s cite only a small proportion of the work cited in those published twenty years previously. Thus, although fields such as physics and chemistry seem to have a stable core of knowledge on which there is substantial consensus, fields such as sociology seem to have a very small core of knowledge with a large research frontier. To the extent that we can use undergraduate textbooks as indicators of the content of the core of knowledge, it appears that the hypothesized differences among the various sciences do exist in the core.

Interpretation of Findings

Can the failure to confirm the hierarchy hypothesis be traced to methodological inadequacies in the study designs? Perhaps the relevant aspects of cognitive structure were not properly measured. Although some sociologists of science would still take this position, I no longer believe that the failure to find systematic differences in consensus among the various scientific fields is a matter of measurement difficulty. Particularly convincing were the data from the peer review study. These data were not responses to questionnaires; they were based on the actual evaluations made by working scientists in the course of carrying out one of their professional responsibilities. That there was little difference in the level of consensus between reviewers in the natural sciences and those in the social sciences lends strong support to the prior evidence collected from surveys and the citation index.

Let us consider two possible critiques of my interpretation of the peer review data. The first is that consensus might be created by the program director, who might send proposals only to reviewers whose approach to the problem was roughly the same as that of the principal investigator. Thus it might be argued that, although there is greater *potential* for disagreement in the social sciences, this disagreement was avoided at the NSF because reviewers were selected only from among those who approach the problem in the same way as the applicant. As an example, consider a proposal from a psychologist to study depression. We might expect that neo-Freudian clinical psychologists and experimental psychologists would have sharply differing views on how to approach this problem. If the proposal was from an experimentalist and was reviewed only by experimentalists, the potential for disagreement would not show up.

This criticism brings up once again the question of the appropriate unit of analysis. From a strictly cognitive point of view it might make sense to define as the unit of analysis all work being done on a particular cognitive problem. Indeed, a traditional intellectual historian might do this. Thus all people studying depression or cancer causation might be grouped together, even if they did not define themselves as being part of the same community and had little or no contact with one another or even awareness of each other's work. Or we might take as the unit of analysis a community of scientists who identify themselves as such and who interact and are familiar with each other's work. In the postscript to the second edition of *The Structure of Scientific Revolutions,* Kuhn (1970) argues that the unit of analysis question caused some of the confusion generated by his book and that his concern was with scientific communities: "Both normal science and revolutions are . . . community-based activities. To discover and analyze them, one must first unravel the changing community structure of the sciences over time. A paradigm governs, in the first instance, not a subject matter but rather a group of practitioners. Any study of paradigm-directed or of paradigm-shattering research must begin by locating the responsible group or groups" (pp. 179–180).

From a sociological point of view, a scientific community is a more appropriate unit of analysis than an artificially constructed group of people created by a researcher who judges them to be working on the same problem.[25] The first is a real social group and the second an intellectual construct. Clinical psychologists and experimental psychologists form two different scientific communities, although they sometimes study the same subject matter. The conclusions I have drawn about levels of consensus are meant to apply only to existing scientific communities. Finally, on this point, the reader should recall that the Phase II experiment suggested that there was no difference in the way in which the NSF program directors and the COSPUP reviewer selectors chose reviewers for particular proposals. Given that COSPUP reviewer selectors made no conscious attempt to send proposals to only like-minded reviewers, this suggests that NSF program directors also were unlikely to have done this. If such a tendency did exist it affected the definition of who was an appropriate reviewer for selectors of COSPUP reviewers as well as for NSF program directors.

The second critique of the peer review study might be that self-selection of applicants serves to create less heterogeneity among social science applicants than exists in the field. In a field such as sociology, for example, many researchers who do quantitative work may apply, but very few ethnomethodologists or historical sociologists. It is probably true that self-selection of this kind does take place. But in cases in which researchers doing different kinds of work do submit grant proposals, reviewers are probably selected from among the relevant scientific community. Thus if more ethnomethodologists and historical sociologists submitted proposals, this change would not necessarily decrease the overall consensus level among reviewers of proposals in the field of sociology. We also do not know the extent to which self-selection based on the style of work takes place in the natural sciences.

Perhaps the most serious problem in this analysis is that we know *how much* scientists in various fields disagree, but we do not know *what* they disagree about. Consider two couples, each of whom is having a theological argument. The first couple is arguing over whether God does or does not exist. The second couple is arguing over the interpretation of some fine point in the Talmud and disagree with each other at least as much as the first couple. But if an outside observer with a broad perspective were to look at these two couples, she would recognize that whereas the first couple disagrees on a very "fundamental" question, the second couple agrees on most "funda-mental" questions but disagrees on much more specific questions. It would be possible for the outside observer to conclude that the second couple is merely being tendentious and arguing over "nothing." This conclusion is equivalent to concluding that the first couple disagrees on a fundamental part of the core whereas the latter couple agrees on the core but disagrees on the frontier. It is possible that disagreement among social scientists might be more akin to the disagreement of the first couple and disagreement among natural scientists frequently more akin to the disagreement of the second couple.

This problem again relates to the unit of analysis. The couple ar-guing over a specific point in the Talmud may perceive themselves as disagreeing as strongly as the couple arguing over the existence of God, because within the confines of what interests each couple there is equal disagreement. It is only when the observer imposes a set of interests which is different from that of the participants that we can conclude that the real disagreement is less for the Talmudic pair or that the real disagreement is less for the natural scientists. From the point of view of how the disagreement affects behavior and social relationships, it might not matter that the scope of the issues on which there is disagreement differs.

Any single conclusion about whether one couple's real disagree-ment is greater than or the same as that of the other couple is impossi-ble. Both answers are correct depending upon the point of view taken. Considering whether there is more agreement between one of these couples than the other is like the problem explored by the fractal geometer Benoit Mandelbrot, when he considered the length of the coast of England (Gleick, 1987). There is no single answer to this question, because it depends upon the detail of the measurement scale. No matter how detailed a scale is used it will always be possible to select a finer scale which will increase the length of the measure-ment. A section of coast which might look smooth from a far distance will look more and more jagged as the distance of observation is reduced.

From the point of view of the sociology of science we should be concerned with the interests of the participants and with what scien-tists believe is important. We must avoid the temptation to be like the observer who tells the Talmudic couple that they are "only" arguing over very minor points. Consider two reviewers of a proposal in solid-state physics. One rates the proposal as "excellent," saying that it represents a brilliant creative solution to a particular problem, and the other rates the proposal as "poor," saying that it is fundamentally flawed and will not succeed in solving the problem. An observer, upon discovering that both reviewers agree that the Bardeen, Cooper, and Schreiffer theory of superconductivity is correct and that quantum mechanics is true, might conclude that the disagreement between them on the proposal is not important. But from the point of view of a sociological understanding of the day-to-day behavior of people as scientists, both the broader agreement *and* the more narrow disagreement are important.

It is because most observers of science have not taken a sociological position which emphasizes the participants' definition of the situation that they have concluded that the natural sciences exhibit higher levels of consensus at the frontier than do the social sciences. In the natural sciences there would also appear to be greater contrasts between core knowledge and frontier knowledge than exists in the social sciences. Failure to recognize the sharp difference between core knowledge and frontier knowledge in the natural sciences contributes to the acceptance of the hierarchy hypothesis.

Conclusion

All the studies that I have made have failed to confirm the hypothesis that we would find higher levels of cognitive consensus at the frontier among the more highly developed sciences at the top of the hierarchy than we would among the less developed social sciences at the bottom of the hierarchy. The structure of knowledge at the frontier is not similar to the structure of knowledge at the core. Whereas the core contains a relatively small number of theories or exemplars, the frontier contains a much broader and more loosely woven web of knowledge. Science at the research frontier is a great deal less ordered and predictable than the traditional view suggests, and this generalization applies to the natural sciences as well as to the social sciences. We have seen that there are substantial levels of disagreement about who is doing important work, what are important problems, and what research should be funded. Because of the great difficulty of reaching consensus on what is good science, reviewer dissensus creates a situation in which whether an applicant to the NSF receives a grant is 50 percent a result of chance—what I have called the applicant's "luck" in the program director's choice of reviewers (see Chapter 4).

The study of consensus is one of the most strategic research sites for the sociology of science. In the development and maintenance of consensus we can see how cognitive factors and social factors interact to create a body of commonly accepted knowledge. Traditional positivists see consensus as ultimately the result of nature. Correspondence with the laws of nature is the primary criterion determining whether scientists accept or reject a contribution. Because scientists have rules which enable them to evaluate the truth content of a contribution, they will be able to reach a consensus. Social constructivists argue that nature or "truth" is irrelevant in determining what comes to be accepted as scientific "fact." What scientists decide to accept or reject is totally a result of social processes. My own position is somewhere between these two. As the evidence presented in this book makes clear, it is not always easy (and not always hard) to determine whether a contribution at the research frontier should be accepted. Clearly there is less consensus in the natural sciences over research frontier knowledge than would be consistent with a positivist view of science. Studying how consensus is formed and maintained is the best way to understand how social processes affect the cognitive content of science at any given time. The failure to find that the characteristics of the phenomena studied influence the ease of consensus formation suggests that broad macro-level characteristics of the cognitive content of knowledge may not have a significant influence on the development of consensus.

6

Evaluation and the
Characteristics of Scientists

In considering how the scientific community evaluates new contributions and reaches consensus that a small number are important and true, I have thus far concentrated on how characteristics of the contributions themselves, as socially perceived, can influence the outcome.[1] But a completely different approach to this question, which will immediately occur to those looking for social influences on the cognitive content of science, is to ask whether the work of some scientists is more readily accepted because they have higher status within the scientific community.

This question was one of those addressed by Mertonian sociologists of science during the 1960s and 1970s. Their research was aimed at examining whether science actually lived up to the ideal norm of universalism, which said that evaluation *should* be based solely upon the quality of the contribution and should *not* be influenced by characteristics of the author. The research concentrated on the problem of accumulative advantage, first introduced by Merton in 1942. Is the work of scientists who are socially advantaged (by being located at a prestigious institution, for example) more likely to receive recognition independent of the quality of the work? Merton's (1968) theory of the Matthew Effect is perhaps the best-known theoretical development of the concept of accumulative advantage.[2]

Merton drew the name of his theory from the Gospel According to St. Matthew, where it is stated, "For unto every one that hath shall be given, and he shall have abundance: but from him that hath not shall be taken away even that which he hath" (25:29). Merton hypothesized that if two scientists of unequal status made the same discovery (an independent multiple), the more eminent of the two would get most of the credit. He extended this analysis to the situation of collaboration between an eminent and a less eminent scientist. From Zuckerman's (1977a) interviews with Nobel laureates, Merton hypothesized that the more eminent collaborator would be given the bulk of the credit even if the less eminent collaborator had done most of the creative work.

In 1970 I published the results of a series of studies aimed at testing the validity of a more generalized version of the Matthew Effect hypothesis. These studies focused on the communication function of the Matthew Effect. Merton had hypothesized that although the misallocation of credit might be harmful at least in the short run for the scientist unfairly deprived of recognition, this misallocation might be beneficial for the communication of scientific information. Because the work of eminent scientists is more likely to be read and used, the misallocation of credit might have the latent function of speeding up scientific communication. I selected random samples of scientific articles which some years after publication received a high number of citations. I then examined the extent to which the articles were cited shortly after publication. I used early (Time 1) citation as the dependent variable, the scientists' status characteristics at the time of publication as the independent variables, and the later (Time 2) citations of the same articles as the quality control. These studies suggested that the quality of the article, as determined by Time 2 citations, was the primary determinant of early recognition and that the characteristics of the authors at the time of publication had only a small independent effect on early recognition. Among papers which were judged to be of equal quality at Time 2, for example, one of which was published by an eminent scientist and the other by a non-eminent scientist, there was not much difference in citation shortly after publication. In another study of papers in all scientific fields, I found that the characteristics of authors of papers which were highly cited at Time 2 but experienced "resistance" at Time 1 were not significantly different from the characteristics of authors of highly cited papers which did not experience delayed recognition. This research suggested that the Matthew Effect did not explain much variance on the evaluation of scientific work in contemporary American science. It seemed that the content of scientific papers was a far more important determinant of their reception than were the characteristics of the authors.[3]

Stewart's (1990) analysis of the plate tectonics revolution in geophysics also reports the result of a quantitative analysis aimed at determining the independent influence of the author's characteristics and the cognitive characteristics of the article on the evaluation of an article. This research, discussed in Chapter 2, also found that cognitive characteristics were about twice as important as characteristics of the author in influencing the evaluation of a sample of geology articles.

Given the emphasis of constructivists on the social determination of evaluation, one would expect them to emphasize the importance of author characteristics. Perhaps because they want to emphasize the social processes involved in the doing of science and to differentiate their work from that of the Mertonians, they have done less work on this problem than one might expect. They do nonetheless frequently suggest that author characteristics play a role in the evaluation of local knowledge outcomes. Knorr-Cetina (1981), for example, suggests that just as the scientists in the laboratory are influenced by their knowledge of how their work will be evaluated, the scientists who do the evaluating are influenced by their knowledge of the individual scientists and of the context that produced the work. Thus who did a piece of work and where it was done influences the way in which it is evaluated.

Data from the NSF peer review study offer us another research site for studying the extent to which the evaluation of science at the research frontier is influenced by characteristics of the scientists. The data will show that when the quality of the proposal is taken into consideration, the characteristics of applicants have very little influence on the way in which their proposals are evaluated. These data support my earlier findings that author characteristics as they have traditionally been studied play only a small role in the evaluation of new scientific knowledge. In the next two chapters I consider the implications of these data and of other research done in the sociology of science for an assessment of whether or not the institution of science operates according to universalistic principles.

Is the NSF Peer Review System an Old Boys' Club?

The NSF peer review study was prompted by substantial criticism of the NSF by members of Congress and others in the 1970s. According to the critics, a group of powerful scientists monopolized grants by giving each other favorable evaluations and excluding from participation younger scientists, those who worked at less prestigious institutions, and others who were not part of the old boys' club. The former Arizona Representative John Conlan charged: "It is an incestuous 'buddy system' that frequently stifles new ideas and scientific breakthroughs, while carving up the multi-million dollar federal research and education pie in a monopoly game of grantsmanship" (National Science Foundation, 1976, p. 40). Criticism of the NSF as an old boys' club is similar to much sociological analysis of the operation of the scientific reward system; such accusations in essence claim that accumulative advantage is unfairly affecting the review process.

Proposals submitted to the NSF represent research as it is created at the frontier. Data from the NSF peer review study enable us to examine at a macro-level how proposals for new ideas are evaluated and thus help us gain insight into the social processes which influence consensus formation.

As I described in Chapter 4, for each application for a grant, an NSF program director generally selects four or five appropriate scientists to act as referees. Each reviewer is sent a copy of the proposal and is asked to evaluate it on the basis of its scientific merit and the ability of the principal investigator.

In Chapter 4 I reported the results of an analysis of variance (fully described in the Appendix) which partitioned all of the variance in ratings received by proposals into two major categories: variance owing to a difference of opinion among equally qualified reviewers and variance owing to the characteristics of the proposal. Roughly two thirds of the variance was a result of reviewer disagreements about the same proposal. One third of the variance could be attributed to characteristics of the proposal.

Here we will look more closely at the portion of the variance which was the result of proposal characteristics. Because reviewers were told to evaluate the proposals on the basis of both the proposal's scientific merit *and* the past track record of the principal investigator, the data reported in Chapter 4 do not enable us to assess the relative importance of these two different criteria. But additional analyses of data from both Phase I and Phase II can be used to assess the relative effect of these two criteria.

Phase I of the research on the NSF peer review system, as we have seen, sought to describe the workings of the system and to determine the validity, if any, of the old boys' club criticisms. Data on 1,200

applicants to the NSF in fiscal year 1975 (120 applicants, half successful and half not, to ten NSF programs),[4] can be analyzed to examine
both how individual reviewers make an evaluation and how the community makes an evaluation of the proposal (the mean rating). To
discover the extent to which the characteristics of applicants influence
the review process, I have used as the unit of analysis individual pairs
of applicants and reviewers. Thus for each applicant I have two to
eight different pairs, depending on how many reviewers there were
for that applicant's proposal. For each of the ten programs I have
approximately 300–500 cases. These enable us to see how characteristics of particular types of applicants are correlated with the ratings
given the proposals by reviewers. All ten programs used the same
rating scale, in which "excellent" was equivalent to 1, "very good" to
2, "good" to 3, "fair" to 4, and "poor" to 5. Some programs allowed
reviewers to give a score between two numbers.[5]

Influence of Applicant Characteristics

According to NSF guidelines for reviewers, the evaluation should be
based on the quality of science described in the proposal *and* the
competence of the principal investigator to conduct the research as
demonstrated by past scientific performance. Data on the reviews
received by the 1,200 applicants of Phase I will allow us to see the
extent to which favorable ratings are more likely to be received by
scientists located at the most prestigious institutions, with the highest
levels of citations and past publication, and by those who have been
successful in receiving NSF support in the past. With knowledge of
the applicant's characteristics, including measures of past research
performance, how well can we predict the ratings that a scientist's
proposal will receive?

The first question is the extent to which scientists who have performed well in the past are more likely to receive favorable reviews
than are scientists who have not.[6] I use three indicators of scientists'
past performance, two based upon citation counts and one upon
number of published papers. The first citation measure is the total
number of citations to the work published by scientists in the last ten
years.[7] Citations are used as a rough indicator of the influence or
"quality" of a scientist's published work. The second citation indicator
includes all 1974 citations to the work of scientists published before
1965. This is a rough measure of the reputations of scientists based

upon work published more than ten years ago. I also examined the numbers of papers that scientists had published in the last ten years. These three measures are direct indicators of the amount of scientific work produced by scientists in the past and the "quality" of that work as judged by the scientists' peers.

We would expect that scientists who have produced the most work in the past and whose work has been the most frequently cited would be the most eminent scientists in their fields. We would also expect that these scientists should get higher ratings than scientists who have produced fewer papers and have been less frequently cited, for two reasons. First, one of the stated criteria of the NSF is the competence of the principal investigators. Presumably scientists who have done the most impressive work in the past should be deemed most competent to do the research proposed in their applications and, therefore, should receive higher ratings. Second, on average, one would expect that scientists who have done the best work in the past will write better proposals for new work.

Let us examine the extent to which ratings may be predicted on the basis of the applicants' characteristics. Consider first the relationship between the number of citations to the recent work of the applicant and the ratings received, using ordinary least-squares regression analysis. Column 1 of Table 6.1 presents the squared zero-order correlation coefficient, which is the proportion of variance in the dependent variable, ratings, explained by the independent variable, citations to recent work. The higher the numbers, the more variance on the ratings can be explained by citations to recent work. In all fields except anthropology, citations to recent work explain some variance in the ratings received. The most interesting fact about these data, however, is that citations to past work explain so little variance in the ratings. Even in biochemistry and chemical dynamics, in which citations explain the most variance in ratings received, they explain less than a fifth of the variance; in most fields, they explain considerably less. Scientists who have demonstrated their competence by publishing frequently cited papers are more likely to receive favorable ratings, but this effect is relatively weak.

After standardizing the data separately within each field I was able to treat all pairs of reviewers and applicants as one sample, in order to consider all ten programs together.[8] When the standardized data for all ten fields are combined, we find that citations made in 1974 to work published between 1965 and 1974 explain only 6 percent of the variance in 1975 ratings of proposals by reviewers.

Table 6.1 Proportion of variance explained (R^2) on rating by characteristics of applicants

Program	(1) Citations to recent work	(2) Citations to old work	(3) No. of papers in last ten years	(4) Recent NSF funding	(5) Rank of current department	(6) Type of current institution	(7) Rank of Ph.D. department	(8) Professional age	(9) Academic rank	(10) All nine characteristics combined
Algebra	.06	.03	.00	.05	.07	.01	.04	.00	.03	.17
Anthropology	.00	.02	.00	.00	.00	.01	.02	.00	.00	.04
Biochemistry	.16	.06	.07	.06	.07	.02	.02	.01	.02	.20
Chemical dynamics	.14	.05	.09	.05	.02	.00	.04	.01	.00	.16
Ecology	.01	.01	.00	.02	.02	.04	.01[a]	.00	.00	.06
Economics	.08	.02	.00	.08	.13	.07	.03	.00	.03	.21
Fluid mechanics	.03	.07	.01	.01	.10	.01	.00	.01	.01	.17
Geophysics	.07	.02	.04	.01	.03	.04	.02	.01	.00	.09
Meteorology	.08	.01	.12	.02	.05	.02	.02	.00	.00	.14
Solid-state physics	.08	.03	.04	.06	.08	.01	.02	.03	.03	.17

a. Relationship is negative.

Column 2 in Table 6.1 presents the proportion of variance on ratings explained by citations made in 1974 to work published prior to 1965. Here we find a positive but very weak relationship in all ten fields, indicating that scientists who are well known as a result of work published more than ten years ago are only slightly more likely to receive higher ratings than scientists who are not well known on the basis of work published ten or more years ago. When I use the standardized data for all ten programs, I find that the citations to older work explain only 2 percent of the variance in rating.

Column 3 of Table 6.1 presents the proportion of variance explained on ratings by the number of papers published in the last ten years. In four of the ten programs (algebra, anthropology, ecology, and economics) no variance is explained. The number of papers published explains only 1 percent of variance in ratings for fluid mechanics, 7 percent for biochemistry, 9 percent for chemical dynamics, and 12 percent for meteorology. When the standardized data for all ten programs are combined, we find that the number of papers published in 1965–1975 explains only 2 percent of the variance in ratings.[9] Our prior expectations as to which scientists would be most likely to get high ratings on their proposals are only weakly supported by the data. If reviewers are being influenced at all by the past performances and reputations of principal investigators, the influence does not seem to be very strong.

I used one other variable as an indicator of the past track records of principal investigators—the number of years out of the last five in which they had received NSF funds. Some applicants had received NSF funds in all or several of the years, whereas others had received NSF funds in none of those five years. The data in Column 4 of Table 6.1 indicate that whether or not the applicants are recent recipients of NSF funds has very little influence on ratings of their current applications. In all ten programs the proportion of variance explained by funding history is low. When the standardized data for all ten programs are used, NSF funding history explains 3 percent of the variance in rating.

These data also tell us whether peer reviewers are more likely to give favorable ratings to scientists in the most prestigious academic departments. We might expect to find such a correlation, given that presumably some departments are more highly ranked than others because they employ more scientists with outstanding reputations. These scientists should get higher ratings because the reviewers

should perceive them more favorably on both of the two stated criteria; they should be perceived as being more qualified and as submitting better research proposals. As the data in Column 5 of Table 6.1
show, however, there is not a strong correlation between the rank of
an applicant's current department and the rating she receives from
peer reviewers. These data lead to the conclusion that reviewers are
not being significantly influenced by the institutional affiliations of
applicants. Applicants from prestigious institutions are only slightly
more likely to receive higher ratings than are those from less prestigious institutions. When we examine the standardized data for all ten
programs combined, we find that rank of current department explains 5 percent of the variance in ratings.

The data presented in Column 6 of Table 6.1 distinguish applicants
currently employed in Ph.D.-granting institutions from those employed elsewhere. This variable has virtually no influence on ratings
of proposals by peer reviewers. Thus the criticisms that the peer
review system unfairly favors applicants from prestigious Ph.D.-
granting institutions are not supported by these data. In Column 7
of Table 6.1 data are presented on the rankings of the departments
in which the applicants earned their doctorates (see Appendix). This
variable also explains very little variance in ratings received.

Some critics of the peer review system hold that young, inexperi-
enced applicants are less likely to receive funds than their older, more
experienced colleagues. An analysis of data on the number of years
since applicants received their doctorates, which I call their professional age, is presented in Column 8 of Table 6.1. In five of the
programs professional age explains no variance in the ratings given
by peer reviewers. In four programs professional age explained only
1 percent of the variance in ratings. These data strongly suggest that
young people are as likely to receive favorable ratings of their proposals as are their older, more experienced colleagues. This conclusion
is supported by the data reported in Column 9 of Table 6.1, which
shows the influence of academic rank (only for those employed in
academic institutions) of applicants on ratings received. A high correlation would indicate that applicants with high academic rank have a
better chance of getting favorable ratings than applicants of lower
rank. Once again, the proportions of explained variance are either
nonexistent or very small. Apparently, full professors do not have a
significantly better chance than their lower-ranked colleagues.

In Column 10 of Table 6.1, using multiple regression analysis, I show the amount of variance explained in ratings by all nine of the applicant characteristics. The characteristics of principal investigators on whom we have data explain only a small portion of the variance in ratings in all ten programs.

Economics is the program in which the largest proportion of variance in ratings—21 percent—is explained by the combination of nine characteristics of the principal investigators. Only 4 percent of the variance is explained by the combination of nine characteristics for the anthropology program. When the data from all ten programs are combined we find that the nine characteristics explain 11 percent of the variance.[10] Because we already know that there is a high level of disagreement among reviewers of the same proposal, it is not surprising to find low levels of variance explained in Table 6.1. But given the fact that small differences in ratings received can determine whether an applicant's grant is approved or denied funding, even the small amounts of variance explained by applicant characteristics may be significant.

Communal Evaluations of Proposals

How might characteristics of the applicant influence the communal evaluation of a proposal? Consider the mean rating received by each of the 50 proposals in each of the three programs in Phase II of the NSF peer review study (see Table 6.2).[11] These correlations, as we would expect, are substantially larger than those when the individual

Table 6.2 Correlation coefficients of applicant characteristics and mean ratings

Applicant characteristic	Program		
	Chemical dynamics	Economics	Solid-state physics
Citations to recent work	.53	.46	.34
Citations to old work	.33	.22	.29
No. of papers in last ten years	.41	.38	.24
Recent NSF funding	.39	.38	.43
Rank of current department	.40	.48	.33
Rank of Ph.D. department	.42	.10	.22
Professional age	.10	−.14	−.15
Academic rank	.26	.30	.06
N cases	(50)	(50)	(50)

rating is the dependent variable. Because we only have 50 proposals for each program, there are too few cases and too many variables to conduct multiple regression analysis.[12] But when we performed a similar analysis on the Phase I data, where we had 100 cases for each program, we found similar correlations to those reported in Table 6.2 (S. Cole, Rubin, J. R. Cole, 1978, p. 84).[13] We may conclude that if we had larger samples the eight applicant characteristics would explain a significant amount of variance on the mean ratings.

The data in Table 6.2, however, are only of limited use in addressing the question of the relative influence of proposal quality and applicant characteristics on the evaluation of the proposals, because we cannot determine whether the correlation between reviewer characteristics and mean ratings is a direct effect of reviewer characteristics or an indirect effect through proposal quality. The only way to separate the direct from the indirect effect would be to have an independent measure of the "quality" of the proposal.

There are some additional data which allow us to make at least an approximate test of the extent to which the applicant characteristics have an independent influence on the ratings received. As I reported in Chapter 4, the Phase II proposals were reviewed under two conditions, nonblinded and blinded.[14] Reviewers of blinded proposals were asked to identify authors of proposals if they could, but were instructed to evaluate the proposal strictly in terms of the content of the science contained in it. If they thought they knew the principal investigator's identity, they were asked to ignore his or her past track record. We can use the evaluations of the "blinded" reviewers as a very rough approximation of an independent assessment of the quality of the proposals.[15]

Because the blinding procedure was not completely successful, we cannot use the blinded ratings of the proposals as a completely independent measure of the reviewers' evaluation of their quality. It would be naive to assume that those reviewers who were able to identify the author(s) of the proposals followed the instructions to disregard this knowledge in making their evaluations. Even though reviewers may have wanted to follow these instructions, their impressions of the quality of the proposal could have been subconsciously influenced by their knowledge of the author(s).

I excluded the reviewers who had successfully guessed who the principal investigator(s) was and computed a mean for each proposal on the basis of those reviewers for which the blinding had been successful. There were some proposals (seven in chemical dynamics, seven in economics, and six in solid-state physics) for which there were no successful blind reviews, and these proposals had to be excluded from the analysis.

The most serious problem in this analysis was the reliability of the "quality" measure. For each proposal, between one and six blinded reviewers were unable to guess the identity of the principal investigator(s). There is substantial disagreement among reviewers, however, and therefore the small sample size of reviewers for each proposal could yield mean quality scores that are unreliable. Because of this problem the results reported below must be considered only tentative.

To get an independent measure of the extent to which the applicant characteristics influenced the mean rating of the proposal, I subtracted the mean of the blinded reviewers from the mean of the nonblinded reviewers. If applicant characteristics have an independent effect on mean ratings, we would expect to see this difference score positively correlated with the applicant characteristics. That is, we would expect the nonblinded reviewers to give proposals of eminent investigators higher ratings than would the blinded reviewers (and perhaps give proposals of noneminent investigators lower ratings than would the blinded reviewers).

Because the number of proposals for which we had blinded means in each program was too small to conduct the analysis separately for each program, I combined the data for the three programs. This provided a sample of 130 proposals for which blind means were available.[16] Because there was a lot of multicollinearity between many of the independent variables and because I wanted to keep the number of independent variables as small as possible, I combined the two citation measures and the productivity measure into a single "quality of past work" index by adding the logged scores for total papers published, total number of recent citations, and total number of citations to older work. I also used academic rank, but not professional age, as an independent variable. The other independent variables were rank of doctoral department, rank of current department, and the number of years in the last five in which the applicant had received NSF support (see Table 6.3).

The only attribute of the principal investigator which has any influence on the mean ratings *independent* of the quality of the proposal is the quality of the applicant's past work as measured by an index combining number of publications and citations to these publications.

Table 6.3 Regression results for predicting difference between means of nonblinded and blinded reviewers related to applicant characteristics

Independent variables	Dependent variable (mean of non-blinded reviewers minus mean of blinded reviewers)
	Beta
Quality of past work index	.35*
Recent NSF funding	.03
Rank of current department	−.01
Rank of Ph.D. department	−.05
Academic rank	−.01
R^2	·.10
Adjusted R^2	.06
N cases	(130)

*Significant at the .005 level.

Even this variable, although the relationship is statistically significant, explains a relatively small amount of variance on the difference score. None of the other characteristics of the applicant, including the rank of the applicant's department, the rank of her doctoral department, her academic rank, or the number of years she had been funded by NSF out of the last five, had any independent influence on the difference score. All the independent variables together explain less than 10 percent of the variance on the difference score. These data provide strong support for the conclusion that the characteristics of the applicant have very little impact on how their proposals are evaluated.

These findings are particularly surprising because reviewers are explicitly told to use the past track record of applicants as one criterion in making their evaluations. Reviewers do give applicants some small amount of extra credit for having done good work in the past, but their overall evaluations are based almost completely on the perceived quality of the proposal.

Influence of Reviewer Characteristics

Finally, there is the question of whether the characteristics of the reviewers have any influence on the evaluation process. Do the status characteristics of reviewers and applicants interact in the evaluation process? Are reviewers from high-ranked institutions more likely to give a favorable evaluation to proposals submitted by colleagues from high-ranked institutions than would reviewers from low-ranked institutions? Data from Phase I were used to test the hypothesis that the way in which evaluation is done in science is influenced by interaction between the characteristics of the evaluated and the evaluators. For each of the ten programs that were studied in detail, I analyzed data on the characteristics of principal investigators and reviewers and on numerical ratings given.[17] Reviewers were significantly more likely than applicants to come from top-ranked departments. In algebra, for example, less than one fifth of applicants and slightly more than two fifths of reviewers came from the fifteen top-ranked mathematics departments. These data by themselves, however, do not tell us anything about how the interaction of reviewer and applicant characteristics might affect the evaluation process. That reviewers are drawn heavily from prestigious departments might only reflect the fact that the most expert reviewers are more likely to be in such departments. Many studies have shown that mean faculty prestige and productivity are highly correlated with departmental prestige (S. Cole and Zuckerman, 1976). Program directors seek reviews from the "best" people in the field, and these people tend to be disproportionately located in top-ranked departments.

In order to see whether the characteristics of reviewers interact with those of applicants, we must determine whether reviewers from higher-ranked departments are more likely to give favorable evaluations to proposals from applicants from higher-ranked departments than are reviewers from lower-ranked departments. An analysis of variance showed that of the ten fields, only in biochemistry are reviewers from top-ranked departments more likely to give favorable ratings to applicants from top-ranked departments than would be expected by chance. This finding could indicate some degree of bias in the reviewing process of biochemistry. In seven of the programs, the relationship had effects opposite to those expected: top-ranked reviewers gave lower scores to proposals from top-ranked applicants than would be expected by chance. These effects were statistically significant, but quite small. In the two other programs, anthropology and meteorology, the differences were not statistically significant.

On the basis of these data, there is very little evidence that the old boys' club hypothesis is correct. The data offer more support for a "jealousy" hypothesis. Because in seven of the ten fields reviewers from top-ranked departments were more critical of proposals submitted by colleagues from top-ranked departments than could be expected by chance, perhaps reviewers are motivated to use slightly higher standards in evaluating the work of those who might be perceived as being their competitors.

For one field, biochemistry, data were collected on the citations of the reviewers. No correlation was found between numbers of citations of reviewers and numbers of citations of applicants. The program director was not more likely to assign the proposals of eminent biochemists for review by other eminent biochemists, at least as eminence is measured by citations. There was also no evidence that reviewers with many citations were more likely to be lenient than were reviewers with fewer citations. As expected, however, applicants with relatively large numbers of citations to their recent work, in general, received somewhat more favorable reviews than applicants with relatively few citations. Most important, the interaction effect was statistically insignificant. Thus no evidence was found to indicate that highly cited reviewers are excessively favorable or excessively unfavorable to the proposals of highly cited applicants. It does not seem that reviews of proposals are influenced to any large extent by an interaction of reviewer and applicant characteristics.

Contradictions between Quantitative and Qualitative Data

In addition to the quantitative analyses reported here, the NSF peer review study included a qualitative analysis of the content of reviews for 250 of the Phase I proposals. Many statements in these reviews suggested that the reviewers were following NSF instructions and giving the past track records of the principal investigator(s) substantial weight in making their ratings. Consider the following excerpts:

> The quality of work, the insight, and the general productivity of programs headed by [PI] are, in my judgment, unsurpassed in the nation. The methodology which they have developed and the kinds of questions which their research addresses are models for the rest of the world to follow . . . In summary I give my highest recommendation for the continued support on this pioneering research group. (rating: 1.0)

> The previous research of [PI] indicates that he could conduct this research program in an excellent manner. Based upon my personal observations, I rate [PI] as being possibly the most competent experimental scientist in this research area. (rating: 1.0)

Although I do not find the proposal to be particularly well written—particularly in terms of a sufficient explanation of experimental design and techniques to be employed, I would still strongly endorse funding of this proposal. This endorsement is based primarily upon the excellent work which has emanated from this group in the field of ———. I have carefully followed the progress of these people over the past several years, and I am rather well convinced that some of the most incisive and yet technically advanced work in the area of ——— in the nation has been performed by this group . . . Had this proposal come from most any other group in the country I would not provide the same enthusiastic endorsement on what is specifically contained in the proposal. (rating: 2.0)

Dozens of similar comments appeared in the reviews examined. The qualitative data drawn from interviews with program directors and panel members as well as the actual statements made in the reviews suggest that the track records of particular principal investigators explicitly enter the peer evaluations and have an important effect on funding decisions. As the quantitative data have shown, however, a conclusion on the importance of these characteristics based upon the qualitative data would be in error. The systematic quantitative data suggest that only a small portion of the variation in reviews is caused by the reviewers' taking into account characteristics of the applicant.

How can we explain the contradictory conclusions drawn from the qualitative and the quantitative data? One possible explanation of this apparent discrepancy is that referees first read and evaluate the contents of scientific proposals on their merits. They decide whether they should be funded or not and then construct arguments to support their decisions and justify their ratings. They comment on the technical science and then, because of explicit instructions from the NSF, they refer to the track records of the principal investigators and their ability to carry out the proposed research. This explanation assumes that the basic adjectival evaluations are made on the contents, the proposals are given tentative ratings, and then the qualitative comments are constructed to conform to NSF criteria and to justify the quantitative evaluation of the proposal. In justifying ratings it is certainly easier and less time consuming to make evaluative comments on applicants' qualifications than to make substantive comments on the contents of proposals. Comments about the past track record of the principal investigator were more likely to be made in positive reviews than in negative reviews. Apparently reviewers felt that to justify a negative review it was necessary to explicate faults in the proposal. Positive reviews could be justified by good points in the proposal or by referring to the past track record of the applicant. It was unusual to find a negative review in which the primary criterion discussed in the review was a negative evaluation of the track record of the applicant. Most important here, however, is the point that only systematic quantitative analysis of random samples enables the sociologist of science to determine the extent to which characteristics of scientists affect the evaluation of their work. Qualitative data is not sufficient to make such a determination.

Explaining the Negative Findings

From the point of view of understanding how evaluation takes place and consensus is created, this chapter's findings are negative. I have identified three types of variables influencing the evaluation process: characteristics of the knowledge itself, characteristics of the scientists, and the operation of intellectual authority. Past research as well as the research presented here suggest that characteristics of the scientists, in so far as these have traditionally been defined, probably have very little influence on the way in which new knowledge is evaluated. Understanding why we obtained these negative findings should enable us to put these results in some perspective.

We found lower correlations between applicant characteristics and individual reviewer ratings for two reasons: self-selection of applicants and lack of consensus among reviewers. The latter has already been discussed in some detail. Because of the lack of consensus among individual reviewers of the same proposal, it was virtually impossible to find high correlations between applicant characteristics and individual ratings. This problem, of course, did not exist in the analysis of the mean ratings. And there we did, indeed, find larger correlations.

Self-selection of applicants to the NSF also caused the correlation between their characteristics and their proposal ratings to be lower than we expected. Most studies of peer review have concentrated on the process by which decisions are made by experts to distinguish who among a group of applicants for limited resources should receive them. Little attention is paid to the equally important process of self-selection, which consists of all the factors that influence the decision of a scientist to apply for a research grant. The applicants for NSF funds are a highly self-selected group, greatly overrepresenting the more active and creative members of the research community.

The proportion of "eligible" scientists applying for NSF funds, of course, varies from field to field depending upon several factors. One of the most important is the availability of other sources of research support. In mathematics, for instance, there is apparently very little support available from other governmental agencies, and therefore a relatively high proportion of university-based mathematicians apply for funds from the NSF. But a relatively small proportion of all scientists employed in Ph.D.-granting institutions in the United States actually apply for research funds.

Consider several pieces of quantitative data related to the self-selection of applicants for NSF research funds. We may compare the distribution of citations to NSF applicants with the distribution of citations to random samples of scientists employed in Ph.D.-granting institutions, specifically, the distribution of citations in 1971 to work published between 1965 and 1969 for physicists, chemists, and geologists employed in the 1971 American Council on Education (ACE)–rated graduate departments.[18]

In 1971 physicists in the rated departments received, on average, 12 citations to work they had published between 1965 and 1969. Twenty-two percent of them had 12 or more citations to their recent work. Applicants to the NSF solid-state physics program in 1975 had a mean of 14 citations to work published between 1970 and 1974. Thirty-three percent had 12 or more citations. This difference between the NSF applicants and the random sample may also be smaller than the real difference between applicants and nonapplicants. First, the random sample undoubtedly contains many scientists who are NSF applicants, and thus the sample does not fully reveal the difference between applicants and nonapplicants. Second, the sample of physicists located at ACE-ranked Ph.D.-granting institutions is itself a relatively elite group and thus does not provide a comparison between NSF applicants and all eligible nonapplicants. Third, solid-state physicists receive somewhat fewer citations, on average, than do physicists in other specialties, such as elementary particle physicists.

The random sample of university chemists had a mean of 14 citations, and 24 percent had 14 or more. The applicants to the chemical dynamics program at NSF had a mean of 20 citations, and 46 percent had 14 or more. Finally, the random sample of university geologists had a mean of 5 citations, and 27 percent had 5 or more, whereas the applicants to the NSF geophysics program had a mean of 10 citations, and 48 percent had 5 or more. These comparisons provide rough and tentative evidence that NSF applicants are a self-selected group of relatively productive and creative scientists.

Understanding that applicants for NSF funds are a self-selected group helps us put in perspective the meaning of the anomalous findings. Consider the finding that there was a relatively low correlation between indicators of past track record and peer review ratings. That a substantial majority of NSF applicants have strong track records means that we have a restriction of range on the independent variables. When the range of variables is truncated, it is difficult to find high correlation coefficients. If scientists with very differing past track records were to apply for funds, we might find a considerably higher correlation between rating received and indicators of past performance.

Consider how self-selection might have influenced the observed correlation between rank of applicant's department and rating received. Although it is true that, on average, highly prestigious departments have more productive and talented scientists, a non-negligible proportion of talented scientists are not in the most prestigious departments. Several independent studies have found that the correlation between citations to a scientist's work and the prestige rank of his or her department is .30 or less (S. Cole and Zuckerman, 1976), indicating that quite a few scientists who have produced high-quality work are not employed in highly ranked departments. When we relate this low correlation to the concept of self-selection, we can better understand the low correlation between the rank of an applicant's department and peer review ratings. If every scientist in every department applied for a grant, there would probably be a considerably higher correlation between rank of department and rating. But we know that all scientists do not apply. Applying scientists from low-ranked departments are probably the most active and creative scientists at these departments. Six mathematicians from a higher-ranked department may apply for NSF funds in a given year, and perhaps only one mathematician at a lower-ranked department will apply. But this one mathematician may have a national reputation comparable to those of some of his colleagues at higher-ranked departments. The relatively wide dispersion of scientific talent and the process of self-selection provide a possible explanation of the anomalous results of the peer review study.

Conclusion

In the past the problem of understanding the social processes that influence the evaluation of new knowledge has not received much attention, because it has been assumed that nature is the arbiter of scientific consensus. The way in which scientists think about and describe core science has confirmed this assumption. We now know, however, that there is no set of objective criteria which enables the scientific community to evaluate each piece of new knowledge and pass judgment on it. Studying how such evaluations are actually made will tell us how consensus is formed and maintained in science.

In order to analyze the process of evaluation systematically I presented data from a large study of how peer review works at the NSF. This study led to the conclusion that even though reviewers were specifically instructed to base their evaluations on the past performance of the scientists submitting the proposals, characteristics of the applicants had very little independent effect on the way in which their proposals were evaluated. These data suggest that, contrary to much sociological theorizing, the status characteristics of scientists in the contemporary United States probably play little role in how their work is evaluated by colleagues.

7

Is Science Universalistic?

The specific content of communal scientific knowledge cannot be fully explained by social factors.[1] Thus sociologists cannot show why the chemical structure of TRF, as accepted by the community of neuroendocrinologists, came to be one sequence of peptides rather than another. Social processes and chance factors can influence the production of new contributions in the laboratory, but as we move from the laboratory into the community local contextual factors play less of a role in the evaluation process. Whereas social constructivists have concentrated on the production of new knowledge, I have focused on the evaluation process. What determines the way in which new scientific contributions are evaluated? We have been considering three types of variables which influence the evaluation of scientific contributions: the cognitive characteristics of the contributions, the social characteristics of the scientists making the contributions, and the operation of social processes such as intellectual authority. This and the following chapter have as their goal the analysis of how social factors as opposed to cognitive factors influence evaluation. The much-debated question of the extent to which the scientific reward system operates according to the ideal norm of universalism provides a vehicle for this analysis.

According to the traditional view of science, the behavior of scientists was supposed to be governed by a set of rules which would ensure objectivity, consensus, and ultimately progress. Social forces could only serve to interfere with the operation of these rules. If science was not universalistic, then progress would be impeded.

Evidence from the NSF peer review study suggests that the characteristics of applicants for grants have a relatively small influence on the way in which their work is evaluated. Thus, for example, eminent scientists from prestigious universities do not have significantly better chances of having their research proposals favorably reviewed than do young scientists from less prestigious universities. Do these and similar findings mean that we may conclude that science is universalistic in the way in which scientific work and scientists are evaluated? In 1973, when Jonathan R. Cole and I published our monograph *Social Stratification in Science,* I would have answered affirmatively. But developments since then in the history, philosophy, and sociology of science suggest that this conclusion was premature, based on both an overly narrow conceptualization of universalism and an overly narrow mode of investigating it empirically. It now appears that the evaluation of new work is influenced by a complex interaction between cognitive factors, such as evidence, models, and theories, and *particularistic* social processes. Frequently it is only possible to make analytic as opposed to empirical distinctions between these two sets of influences. And perhaps most important, at the micro-level, when we consider the evaluation of the work of most scientists on a day-to-day basis, the particularistic processes seem to dominate. The high state of underdetermination makes the universalistic processes operate at a remote distance. By this I mean that in the complex evolutionary process by which contributions survive or die, the extent to which they match the empirical world is clearly a driving force. But at the time most contributions are made it is difficult or impossible to determine this.

Theoretically I accept the position of the constructivists and of many contemporary historians and philosophers of science that there is no way to evaluate *objectively* most knowledge produced at the research frontier. If there are no predetermined objective criteria, then all evaluation of frontier science is inherently subjective and therefore particularistic. There remains, of course, the significant question of in what ways particularistic criteria are being employed. Prior studies in the sociology of science have suggested that particularism based upon the nonscientific statuses of scientists (such as religion or race) is currently not widespread. Other studies of particularism based upon scientific affiliations (such as the prestige of one's doctoral department) show that such affiliations have some (but, I argue, a limited) impact. But sociologists have failed to study the primary basis for particularism, the location of scientists in networks of interpersonal connections. Network ties based upon personal, social, and cognitive factors have a significant influence on how the work of scientists is evaluated. Why has past research failed to find much evidence for particularism when all of us are familiar with many examples from our own experience?

Before seeking an answer to this question, it is important to note the relationship between the evaluation of scientific contributions and the evaluation of the scientists who have made these contributions. In the day-to-day doing of science it is usually impossible to separate the evaluation of scientists and the evaluation of their work. There are strong norms in the scientific community specifying that the most important criterion on which scientists *should* be evaluated is the "quality" of their intellectual contribution. Although there may be many scientists who believe that this ideal is frequently not attained, there are very few who would not agree that this norm *should* be followed. But when a new contribution is evaluated, its author is also being evaluated. And most important for the focus of this chapter, when an individual scientist is being evaluated, that scientist's contributions are also being evaluated. When a scientist is rewarded, that individual's prestige is increased *and* the prestige of that individual's contributions to science is increased. The evaluation of individual scientists, therefore, is a crucial part of the evaluation of the contributions they have produced. Consider, for example, a young woman who is offered tenure at a highly prestigious university on the basis of a book that derives from her doctoral dissertation. This reward increases the young woman's prestige, but it also increases the prestige of the book. Other scientists who might not have paid attention to the book now come to perceive it in a favorable light. Some of these scientists may have actually read the book, and others will accept it as being a "significant" contribution without reading it. Or consider a young scientist who has been working on an innovative research program for a number of years and then is denied tenure by the prestigious department in which he works. This decision influences how the community perceives the scientist *and* how it perceives the research program he has been working on. In the analysis that follows I frequently look at how individual scientists are evaluated, but we should not lose sight of the fact that the evaluations of scientists play a crucial role in determining what happens to the contributions the scientists have made.

The Mertonian Norm of Universalism

In an early essay on the "ethos" of science Merton ([1942] 1957a) listed universalism as the first of four ideal norms which ought to govern the behavior of scientists. He defined scientific universalism as having two components. First, claims to truth should be evaluated using impersonal and objective criteria and should not be influenced by any characteristics of the authors. Second, careers should be "open to talent," that is, everyone should have an equal opportunity to achieve scientific success, with evaluation being based on the quality of role performance. Merton saw the operation of such norms as a functional requirement for the advancement of scientific knowledge. If personal biases and prejudices influenced the evaluation of new contributions, it would take longer for nature to reveal its laws. "The circumstance that scientifically verified formulations refer to objective sequences and correlations militates against all efforts to impose particularistic criteria of validity" (Merton, [1942] 1957a, p. 553). It is implied that the moral norms exist *because* they allow science to receive the necessary evaluation utilizing "pre-established impersonal criteria": "Universalism finds further expression in the demand that careers be open to talents. The rationale is provided by the institutional goal. To restrict scientific careers on grounds other than lack of competence is to prejudice the furtherance of knowledge. Free access to scientific pursuits is a functional imperative. Expediency and morality coincide" (pp. 554–555).

Merton's claim that universalism was an essential part of the ethos of science has probably generated more research and more controversy than any other topic in the sociology of science (Zuckerman, 1988). In the late 1960s many sociologists of science took up the challenge of empirically investigating the extent to which the contemporary scientific community (mostly in the United States) actually realized the ideal norm of universalism (S. Cole and J. R. Cole, 1967; S. Cole, 1970; J. R. Cole and S. Cole, 1972, 1973; Crane, 1965; Hagstrom, 1971, 1974; Hargens and Hagstrom, 1967; Gaston, 1973, 1978; Zuckerman, 1977a). Most of these studies examined the extent to which a particular reward was correlated with measures of role performance, such as the number of papers published or the quality of the papers as indicated by citations.[2] Because the measures of role performance were never more than moderately correlated with measures of rewards, it was still possible that the distribution of rewards would be influenced by nonuniversalistic (particularistic) criteria.

The particularistic criteria studied were divided into two groups: the nonscientific statuses of the scientists, such as age, religion, gender, and race, and the scientific affiliations of the scientists, such as the rank of their Ph.D. department, the rank of their current department, or past receipt of scientific rewards. Particularism, of course, need not be based upon these characteristics but could also be based on a host of other irrelevant personal characteristics, such as physical appearance, manners, personality, or political views. The independent variables were selected for two reasons. First, it was easy to collect data on variables such as gender and rank of Ph.D. department and very difficult to collect data on variables such as personality characteristics or political views. Second, in the sociology of science as in other areas of sociology, research interests (the foci of attention) are molded by what is going on in the society at the time. Discrimination against minority group members and women has been and continues to be a major issue in the larger society and therefore drew the attention of sociologists of science as they studied the operation of the reward system. Discrimination on the basis of race and gender in an earlier historical period had been common practice. Now it was illegal, but was it still being practiced in the scientific community?

The early work done on universalism in science did not establish a consensus. No researchers argued that science was completely universalistic; clearly it was not. But many argued that science seemed to come closer than other institutions to approximating its universalistic ideal. In *Social Stratification in Science* and in later work (S. Cole, 1978; S. Cole, Rubin, and J. R. Cole, 1978; J. R. Cole, 1979), colleagues and I concluded that in contemporary American science rewards were generally (but not always) distributed on the basis of universalistic criteria, and that to a large extent rewards were meted out in accord with demonstrated role performance.

This conclusion, though accepted by many, has been strongly challenged by others. Challenges have come from two sources within the sociology of science. First, American sociologists doing work similar in style challenged our conclusions as being empirically incorrect. We had, according to these critics, underestimated the extent to which irrelevant characteristics such as gender influence the distribution of rewards, and we had misspecified the direction of the relationship between variables such as scientific productivity and rank of academic department. Second, as will be discussed in Chapter 8, British sociologists of science, influenced by a relativist philosophy of science and the new historiography of science, argued that our acceptance of positivistic assumptions caused us to misinterpret the meaning of the empirical results we reported. Given the importance of the evaluation system in the development of systematic cognitive knowledge, it is imperative that our earlier conclusions on universalism be reexamined. In this chapter I will look at the empirical critiques which have been made of my earlier work on universalism by American sociologists of science. In the following chapter I will consider the conceptual critiques made by constructivist sociologists of science.

Rewards and Functionally Irrelevant Statuses

A functionally irrelevant criterion is one which has no bearing on the ability of an individual to perform a particular functional role or status.[3] Nonscientific statuses such as gender and race are functionally irrelevant for the doing of science, and to the extent that they are used as explicit or hidden criteria in the evaluation of scientific work and the distribution of rewards, the principle of universalism is being abridged and discrimination in the workplace is being practiced. There is ample evidence that in the past religion, race, and gender (Rossiter, 1982) made it difficult and sometimes impossible for a student to receive training at some leading graduate departments, let alone faculty membership in those departments. Although a few members of disadvantaged groups pursued scientific careers, they were generally confined to marginal positions (J. R. Cole, 1979).

To what extent do functionally irrelevant statuses play a role in the reward system of contemporary American science? There are two questions which must be addressed. First, are some people prevented from entering science because of discrimination? Second, once people have earned the Ph.D., to what extent is the way in which they are evaluated influenced by functionally irrelevant statuses? Currently in the United States various federal statutes and administrative regulations make it illegal to discriminate in admitting students or in hiring scientists. Most scientific institutions have affirmative action officers and procedures designed to ensure not only that women and minorities have an equal chance of being admitted as students and employed as scientists but that the institution makes a special effort to recruit qualified members of these groups. The extent to which affirmative action programs are taken seriously and are effective, of course, varies from institution to institution and even among departments within the same institution.

The fact that as of 1990 less than 15 percent of scientists were women and less than 5 percent members of minority groups (excluding Asians) has led many social scientists and members of the general public to believe that discrimination is keeping members of these groups out of science. It has been *assumed*, for example, that women have not been afforded the same opportunity as men to enter science. This assumption is based upon the "unequal ratio fallacy," which assumes that anything other than an equal ratio of women and men entering a high-status occupation such as science is evidence for sex discrimination. An outcome is taken as evidence of a process. Discrimination may, of course, explain some, or all, of an unequal sex ratio in the distribution of a particular reward, but an unequal outcome in itself does not measure the effects of discrimination any more than it measures the effects of biological propensities or gender norms. By misinterpreting descriptions of outcomes as indicators of process, the unequal ratio fallacy imposes a single variable causal model that considers only opportunity as a possible cause of the underrepresentation of women and minorities in science (S. Cole and Fiorentine, 1991).

We do not have a full understanding of the variables that influence the career selection of various social groups. We do know that outcomes such as these can be influenced by both self- and social selection processes. The values posited for men and women by gender culture, for example, seem to play a significant role (S. Cole, 1986; Fiorentine, 1987, 1988; Fiorentine and S. Cole, 1992). Much more research is needed to understand the interplay of restricted opportunities and cultural values in influencing the decision to become a scientist. We do have better data on what happens to men and women once they make the decision in college to pursue a graduate education and a career in science. The evidence suggests that today male and female applicants to graduate programs and fellowships have an equal chance (Bickel, Hammel, and O'Connell, 1975; J. R. Cole, 1979).

Among those who have received Ph.D.s in science, my research has indicated that at least from the late 1960s functionally irrelevant nonscientific statuses did not seem to have a strong impact on the distribution of rewards. For example, age is not correlated significantly with visibility, a measure of peer recognition (J. R. Cole and S. Cole, 1973, p. 107); with whether or not a scientist has received an NSF grant (S. Cole, Rubin, and J. R. Cole, 1978); or with the scores a scientist has received from NSF reviewers (Chapter 6). Jonathan Cole and I also showed that among a sample of 300 full professors in Ph.D.-granting departments in five different disciplines Jewish scientists had slightly higher visibility and were in slightly higher ranked departments than were non-Jewish scientists (J. R. Cole and S. Cole, 1973, pp. 154–159). We concluded that religion did not seem to play any role in the evaluation of university scientists.

The functionally irrelevant status which has received the greatest attention in my own work and from other researchers is gender. Women have received less recognition than men in science. Whether we look at the proportion of Nobel Prize winners or National Academy members or recipients of any other prize who are women, we find women underrepresented. How can we explain the correlation between gender and recognition in science? Once again, the unequal ratio fallacy is used to assume that the answer is discrimination. If an unequal proportion of men and women have received recognition it is assumed that men and women who have made similar contributions are treated differently or that women have not been afforded the same opportunity men have to make important contributions.

A recent example of the application of the unequal ratio fallacy may be seen in the work on gender discrimination in science by Haberfeld and Shenhav (1990).[4] Using data from two longitudinal surveys of American scientists and engineers, they conducted a regression analysis in which the dependent variable was income, with type of degree, length of work experience, family characteristics, and type of job controlled. When the regression coefficient for gender was found to be statistically significant, the conclusion was reached that gender discrimination was being practiced. Discrimination is possible, but in the absence of data on both role performance and motivational differences it is unwise to conclude that an outcome such as a gender difference in income is a result of a process such as discrimination. Elsewhere Robert Fiorentine and I have illustrated how income differences between men and women scientists can arise from differing behaviors influenced by gender norms without being a result of discrimination on the part of employers (S. Cole and Fiorentine, 1991).

In the most thorough study conducted on women in science, Jonathan R. Cole (1979) found generally low correlations between gender and both the rank of academic department and the receipt of honorific awards. He did find a larger correlation between gender and academic rank, a correlation which remained statistically significant but relatively small after controlling for career interruptions and scientific productivity. The general conclusion reached by us in our early research (J. R. Cole and S. Cole, 1973, chap. 5) and by J. R. Cole (1979) in his more comprehensive analysis was that in most situations in contemporary American science gender plays little role in the distribution of recognition. Where evidence does exist that gender may play a role, it generally explains very little variance.[5] There is no new evidence on functionally irrelevant nonscientific statuses which would cause us to modify the conclusion that these statuses play only a small role in the evaluation of scientific work.

Rewards and Scientific Affiliations

My earlier research on social stratification in science led to the conclusion that there are two primary determinants of receipt of scientific recognition: role performance, as measured by indicators of scientific output, and the process of accumulative advantage. The process of accumulative advantage, as we saw in Chapter 6, refers to the fact that once a scientist has been rewarded, his or her chance of receiving further rewards in the future are greater, *independent* of indicators of role performance. Additional evidence supporting the hypothesis of accumulative advantage has been presented by Allison and Stewart (1974), Zuckerman (1977a) and many others.[6]

As pointed out in J. R. Cole and S. Cole (1973, p. 235), because in a completely universalistic reward system quality of role performance should be the sole criterion upon which scientists are rewarded, the operation of accumulative advantage represents a departure from the ideal. Whether such a departure is "bad" for science depends on the answer to the most important question generated from the study of accumulative advantage: Is this process an outcome of the unequal distribution of talent, which tends to cluster at the prestigious centers, or is "talent" a result of the unequal distribution of resources and facilities (J. R. Cole and S. Cole, 1973, p. 75)? If accumulative advantage serves to allocate more resources to the most talented scientists, it might be unfair to some individuals but beneficial for the general advance of knowledge. But if accumulative advantage causes resources to be unfairly allocated to scientists who cannot make the best use of them, it is harmful for the general advance of knowledge. Because only a small minority of working scientists do work which has a significant impact on the state of knowledge, we reached the tentative conclusion that accumulative advantage could not serve to elevate a nontalented scientist to a position of eminence, but could serve to give greater rewards to a "mediocre" but well-situated scientist over a "mediocre" scientist who was not as well situated. The implication was that it was primarily talented scientists who benefited from accumulative advantage and that the process, though unfair to individuals, probably had a positive impact on the rate of scientific advance.

Long (1978) raised serious questions about this interpretation. In our research we had found, as had others (Crane, 1965; Hargens and Hagstrom, 1967), that the prestige of the doctoral department and measures of scientific productivity were equally correlated with the prestige of a scientist's current department. The absence of longitudinal data made it difficult to interpret the meaning of these correlations. Looking at a cohort of biochemists, Long found that although predoctoral productivity was the single best predictor of productivity ten years later, the quantity and quality of publications prior to the first job were not significantly correlated with the prestige of the first job. The prestige of both the doctoral institution and the institution in which a scientist had a postdoctoral fellowship and the eminence of the mentor did have significant effects on the prestige of the first job. In addition Long showed that the correlation between prestige of department and productivity increases over time and argued that this is a contextual effect in which the prestige of the department influences the productivity level, rather than the reverse. Long (1978) concluded:

> Advantage accumulates not to those who have been successful at time 1, but to those who have received the advantage of a prestigious position for reasons independent of their productivity at time 1. To the extent that the eminence of a scientist's mentor and the prestige of his doctoral department, *independently* of the productivity of the scientist, are particularistic criteria for evaluation, a *particularistic* advantage accumulates, not an advantage initiated for universalistic reasons. (p. 905)

The same data were further analyzed by Long, Allison, and McGinnis (1979). They argued that because pre-employment productivity is not related to the prestige of the scientist's first job but is the strongest predictor of later productivity, science is departing from its universalistic ideal. Science may be hurt, they said, as a result of the initial "misallocation." "The strong effect of pre-employment productivity on later productivity suggests that departments are being nonmeritocratic in ignoring these variables in their hiring decisions" (p. 828). Long and McGinnis (1981) showed that whether scientists are employed in contexts which encourage research or contexts which discourage research has a substantial effect on later productivity independent of initial productivity.

Allison and Long (1987) present an analysis of 274 job changes by academic scientists. They find that prestige of the prior job, prestige of the doctoral department, and quantity of publications (but not quality as measured by citations) influence the prestige of the destination department. In another paper (1990) they show that the prestige of the destination department has an independent effect on future productivity. In these two publications they are considerably more cautious than in earlier ones in interpreting the data as evidence for the absence of universalism and the utilization of particularistic criteria.

The work by Allison, Long, and McGinnis represents a substantial increase in our knowledge about the process through which accumulative advantage works, with an emphasis on how social contexts serve to increase or decrease scientific output. Their work has led to the important conclusion that institutional context has an independent effect on future productivity even when past productivity is taken into account. But their work also leaves unanswered important questions about the meaning of some of the correlations they observe. Is the correlation between prestige of department and measures of output entirely a contextual effect, or a result of "unfair" accumulative advantage as opposed to "fair" accumulative advantage?

In their work on the determinants of prestige of first appointment, Allison, Long, and McGinnis treat correlations between this variable and the two variables "prestige rank of the Ph.D. department" and "prestige of the mentor" as evidence for the operation of particularistic criteria. But they also point out that these last two variables may be used by the hiring departments as indirect indicat ɔ of the *future* scientific contribution of young scientists. There is a danger here of reifying the measures; that prestigious departments end up hiring young scientists who have come from other prestigious departments or who have eminent mentors is not evidence that these have been the *actual* criteria used. The correlations may be spurious, a result of the influence of talent on both the prestige of graduate school and the prestige of the hiring department. Even though this talent may not be correlated with predoctoral productivity, it may be evident in colloquia and other face-to-face interaction between the candidate and members of the hiring department. Allison, Long, and McGinnis show that there is a moderate correlation between both the prestige of the Ph.D. department and the eminence of mentor and later productivity. The independent effects of these variables are reduced when the prestige of the first job is controlled for, but this could indicate that prestige of the first job is an intervening variable specifying the mechanism through which talent leads to later productivity. There is no way to tell whether the correlation at Time 2 between productivity and the prestige of the first job results only from characteristics of the environment or from an interaction between characteristics of the environment and characteristics of the scientists *selectively* recruited.

To deal with this problem, Allison, Long, and McGinnis analyze data on job changes. A scientist's ability is unlikely to change when he or she changes jobs. Changes in productivity, therefore, must be a result of the changed environment. The analysis supports the conclusion that changed environment influences change in productivity. But data on the mechanisms through which this correlation is brought about are lacking. Allison, Long, and McGinnis speculate that scientists in more prestigious contexts may have more access to the resources needed to do research. This speculation is probably true, to some extent, but for several reasons seems unlikely to explain the bulk of the variance due to context. First, we saw in Chapter 6 that among scientists who apply for NSF research grants, rank of department or whether the scientist is from a Ph.D.-granting department or some other type of institution explains very little variance both on reviews and on whether the scientist receives the grant. Second, the strength of the correlation between rank of department and the provision of local research resources is questionable. Some lower-prestige or even unranked departments provide more resources than some high-ranked departments. Third, the correlation between context and citations (not productivity) may be a result of a halo effect. Scientists at more prestigious departments are more visible, and scientists who move to a prestigious department will have their visibility increased as a result of the move. This visibility can substantially increase citations. Fourth, it seems likely that the primary way context influences productivity is by change in motivational level rather than by change in the allocation of resources. Several studies provide empirical evidence showing that scientists who are rewarded will produce more in the future (J. R. Cole and S. Cole, 1973; S. Cole, 1978; Reskin, 1977).

The data Allison, Long, and McGinnis present do not, in my opinion, prove that scientific ability has little influence on productivity. The data on job changes mix people who have changed jobs or decided not to change jobs for many different reasons and therefore do not allow an unambiguous interpretation. The correlations between rank of department and other variables, including output measures and rank of doctoral department, are relatively low. Thus there are scientists with a large number of publications and citations who are employed at departments of relatively low prestige and there are scientists with very few publications and citations who are employed at high-prestige departments. The rank of a scientist's department is a result of the combination of both social selection and self-selection processes.[7]

Departments must choose whom they want to hire, but scientists must also choose where they want to work. This distinction is important in the universalism-particularism debate. If a scientist with high productivity is at a low-prestige department because the high-prestige departments do not want to hire him, his circumstance may be evidence of the use of particularistic criteria. But if the high-productivity scientist is at a low-prestige department because she *chooses* to be there, her situation provides no evidence for the use of particularistic criteria. That a scientist chooses to leave a high-prestige department at, say, the University of Chicago to go to an unranked department at Dartmouth will look like an example of particularism in a statistical analysis such as that conducted by Allison, Long, and McGinnis. That the same scientist then decides to go from Dartmouth back to the University of Chicago and then to Harvard will look like an example of universalism. That a highly respected scientist chooses to turn down offers from several prestigious universities in order to remain at a university of lower prestige will reduce the correlation between rank of destination department and research output but is not relevant for the question of universalism in science.

Consider the data used by Allison and Long (1987) on job changes. Their sample consists only of those people who have actually changed jobs. Suppose that we had data on all scientists who had been *offered* a job at another department and that among this group there was no independent effect of the prestige of the doctoral department on the prestige of the offering department when productivity was controlled.[8] Allison (private communication) suggests that this would be unlikely, because the idiosyncratic factors influencing self-selection should be randomly distributed and therefore reduce but not eliminate significant effects. But what if the value placed on departmental prestige in selecting a job was not randomly distributed? It is quite possible that scientists who were educated at prestigious graduate departments might place a higher value on the prestige of their current department than do those educated at less prestigious departments. In this case the former group would be more likely to accept job offers from prestigious departments, thus creating a spurious correlation in the Allison and Long data. I am suggesting not that this is what is occurring, but only that these data do not enable us to rule out this and other reasonable hypotheses which would explain the outcome without indicating that prestige of doctoral department was the basis of particularistic evaluation.

Because of the difficulty of separating the operation of social selection and self-selection the data presented by Allison, Long, and McGinnis do not enable us to draw strong conclusions about the extent to which institutional origins of scientists form the basis of particularistic evaluation. Because these two social processes are so empirically intertwined, if one wants to assess the independent effect of social selection it is necessary to find a strategic research site where self-selection is eliminated as a possible explanation. A study of *applicants* for scientific jobs could be conducted. Today positions must be posted, and applicants, even those invited to apply, must submit a formal application. If we had data on applicants and on which candidates were offered positions (including those who declined the offer), we could conduct an empirical study of the extent to which universalistic criteria were being applied in the distribution of academic positions. This situation is one in which we would probably find substantial amounts of particularism in individual cases, as I point out below, but it would be useful to know how the system worked at an aggregate level.

Allison, Long, and McGinnis do not analyze data on reputational success, or the reputations of scientists as judged by others. This variable is *not* voluntaristic and is, therefore, not influenced by self-selection. In my analysis of the determinants of "perceived quality" of scientists (a mean ranking obtained from questionnaires) I found both rank of doctoral department and quantity of publications to have relatively small influences, and the quality of publications to have a large influence. Perceived quality of the scientist had a very strong influence on that scientist's visibility, and the rank of current

department had a small independent effect on visibility. These findings were consistent across the five fields studied (S. Cole, 1978).
Thus, for rewards based on recognition, it appears that the work
produced by scientists and the opinion of this work in the scientific
community are by far the most important determinants. Where scientists came from or where they are currently located has significantly
less impact on how they are evaluated.

The research on the peer review system at NSF also avoids the
problem of confounding self-selection and social selection, because it
studies samples of applicants. What criteria influenced the distribution of perhaps the most important resource needed to do science,
money for research? The perceived "quality" of the proposal, rather
than the characteristics of the scientists, was the most important determinant of the scores applicants received from peer reviewers.

Long (1978) concludes: "Overall there is no evidence of the importance of productivity in the recruitment of faculty" (p. 897).[9] Long's
(1978) conclusion does not seem justified on the basis of the evidence
presented by Allison, Long, and McGinnis. The problem may be that
they are too concerned with analyzing variance and may be missing
the underlying uniformity. (See Lieberson's [1985] analysis of how
sociologists might study gravity.) Although it is true that productivity
as a graduate student or postdoctoral student does not seem to explain variance on first job and that some scientists receive internal
promotion without publishing much, it would be difficult to list cases
in which an outside appointment to a Ph.D.-granting department was
made at the tenure level to an appointee who did not have a significant publication record.[10] Given that the overwhelming majority of
scientists do not have significant publication records and will not,
therefore, be eligible for consideration, Long's conclusion seems un-
warranted. A more accurate statement might be that a significant
publication record is a *necessary* condition to be recruited at the tenure
level, but among those who have significant publication records, differences in number of papers and citations may not explain much
variance in the prestige of the department making the appointment.

Despite the important work done by Allison, Long, and McGinnis,
the perceived "quality" of a scientist's work remains in general the
most significant determinant of recognition. The extent to which particularism enters the evaluation system of modern American science
through the utilization of functionally irrelevant nonscientific statuses
and even through the use of what might be debated to be functionally

irrelevant scientific affiliations (for example, the prestige of where one obtained one's doctoral degree) is not very great.

Case Studies of Particularism

Although statistical findings such as those presented in the previous chapter and in my previous work suggest that particularistic evaluation plays only a small role in science, these data do not mesh well with the day-to-day experience that many of us have had in science. In our personal observations of the reward and evaluation systems of science, we see scores of examples of particularism. In this section I give just a few examples which have come to my attention in twenty-five years of participant observation in academic science.[11] Cases or examples do not prove anything, but they do illustrate the kind of events which lead many scientists to question whether the reward system indeed works equitably.

Consider a National Science Foundation program director who receives an application from a colleague to whom he does not want to give a grant. He believes that on intellectual or cognitive criteria this colleague does not deserve to receive a grant. He sends out the proposal to four reviewers, three of whom give the proposal favorable ratings and the fourth a lukewarm rating. The program director, saying that he has more faith in the opinion of the "lukewarm" reviewer than in the other three, rejects the application and even goes so far as to say that he sent the proposal to the other three reviewers more as a test of the reviewers' ability than to find out the merits of the proposal.[12] This case, in which the program director seems to have made up his mind not to give a grant prior to the review process, would seem to be a clear example of particularism. The program director's decision makes it less likely that the ideas of the grant applicant will have an opportunity to be developed and accepted by the community.

Or consider a case from another granting agency. A scientist agrees to help a program director by doing some particularly arduous committee work. Later the scientist has an application for a grant approved by this program director. The program director would claim that the proposal was meritorious. But had the principal investigator not become a "friend" of the program director by doing him substantial favors, it would have been less likely that the proposal would have been funded.

Consider these examples of the hiring process. A biochemistry department wants to hire two assistant professors. They interview about a half a dozen candidates and then discuss their choice. A male candidate is labeled "arrogant," even "obnoxious," but "brilliant." The members of the department who believe that this candidate will "disrupt" the department debate those who believe that his work is so good that his abrasive personality should be overlooked. A decision is made to hire this candidate. The next decision boils down to a choice between two women candidates. Once again one of the two candidates is called "abrasive" and "difficult to get along with." Some members of the department believe that this candidate is engaged in a brilliant research program. Yet the department decides not to hire the abrasive female and to hire a more pleasant but less brilliant woman for the second position. Some members of the department would argue that the work of this second nonabrasive candidate was as good or better than that of the abrasive candidate.

This example offers several possible abridgments of the principle of universalism. First, it shows the great significance of the assessment of "personality" in making certain decisions. Second, there was an unwritten rule in the department that it would have to hire one male and one female; thus, gender was introduced as a criterion in the hiring process. Third, the department to at least some observers was sexist in accepting abrasive behavior from a male candidate yet finding that similar behavior from a female candidate disqualified her. But it should be stressed that some members of the department would reject this interpretation and argue that cognitive grounds were the most important basis of the decision.

Consider another case of an economics department which is hiring an assistant professor. A half a dozen or more candidates are interviewed. There is not strong enthusiasm for any of them, but the one who is offered a job turns out to be a young scientist who has recently published a highly laudatory review of the chairperson's latest book. The chairperson would strongly argue that this scientist was the best candidate.

Or consider a computer science department in which a woman and a member of a minority group are being considered for a tenure decision. If these two scientists had not been either women or minority group members, it is likely that neither one would have been considered for tenure, because on the basis of their publication records they do not meet the university's normal criteria. It is only when "affirmative action" considerations are thrown into the equation that either candidate can be considered. In this case the department recommends that both receive tenure, and the ad hoc committees established to review the cases recommend that neither receive tenure. In one case the university officials deny tenure and the faculty member is forced to leave (but is hired by a department at another prestigious university). In the other case tenure is granted. This decision can be understood in the light of the relationships of the two scientists to other powerful professors in their department and to the most powerful university administrators. The scientist denied tenure has some significant enemies among the faculty and no friends among the administration. The scientist granted tenure has no strong enemies among the faculty and good friends among the administration.

Or consider another tenure case in an ecology department. A bright and productive assistant professor with favorable letters of recommendation and publications in the field's most prestigious journals is denied tenure because he criticized publicly the dissertation of the chairperson's favorite student. This reason for the opposition to his tenure is never openly discussed, but he is criticized for being a "poor colleague" and his work is subjected to an unusually critical and hostile review. The assistant professor is hired as a tenured associate professor at a university of equal prestige and goes on to have a successful career. The friend of the chairperson is hired as an assistant professor, promoted to associate professor, and later advanced to full professor despite a relatively weak publication record.

Rubin (1975) in his analysis of the dynamics through which tenure decisions are made in chemistry and sociology describes a sociology department in which the criterion for tenure is to publish an "important" monograph. He interviewed four individuals who had received tenure and three who had not: only one of the seven had published an important monograph. "Out of the remaining six, five individuals had very similar publication records (on the basis of their vitae), but only two of them received tenure. And one individual who received tenure had only one article to his credit, significantly less than everyone else" (p. 247). He goes on to describe a situation in which factors such as political connections with senior faculty members, personal life style, and whether or not one entertained were the determinants of whether or not tenure was awarded (pp. 246–251). Similar factors were frequently mentioned by subjects in Caplow and McGee's (1958) classic study, *The Academic Marketplace*.

Other examples of particularism in hiring may be found in the increasingly widespread practice of "package deals." Universities used to have nepotism rules, which prevented members of the same family from being employed in the same department. With the increasing frequency of "dual career" couples, these rules have all but been eliminated. It is not unusual for a university to offer a high-paying tenured job to the less eminent spouse of an eminent scientist in order to attract the eminent scientist. When the package offer is made, the department must argue that both candidates are deserving of job offers in their own right. But virtually everyone knows that the less eminent member of the package would never even have been considered for the job were it not for his or her attachment to the more eminent part of the package.

Particularism can also frequently be seen in the distribution of awards, prizes, and fellowships. An eminent scientist serves on a committee to decide how a foundation's funds should be spent and then is one of the first recipients of the prize the committee decided to distribute to deserving scientists. Several scientists are asked to form an outside committee to recommend candidates for a prestigious department that has been having a particularly hard time recruiting. Leading students of some of the members of this committee are then recommended for the jobs. Students and friends of eminent members of committees charged with giving out scarce rewards and fellowships are much more likely to receive these awards. A scientist assigned to a committee to select the most important scientific contribution in a certain area vows to prevent an enemy who has been nominated from receiving the award and succeeds in so doing; because there are many qualified candidates, strong opposition to anyone is often enough to prevent that individual from receiving the award.

The above represent only a very small unsystematic sample of cases of particularism I have come across. Many readers of this chapter, I am sure, can add cases from their own research or personal experience.[13] Such cases pose an important paradox. The research which has been conducted at an aggregate level has not produced any conclusive evidence that particularism plays more than a small role in the way in which scientists and scientific work are evaluated. Yet at the day-to-day individual level, science, like other institutions, seems to be riddled with particularism.

Conclusion

Statistical studies of universalism have generally shown particularistic variables to explain only a small amount of variance in rewards in part because these studies ignore the most important bases of particularism. We have defined particularism as occurring when scientists with particular statuses or affiliations (such as young people, women, people who were trained at prestigious departments) receive greater or lesser rewards than can be explained on the basis of the quality and quantity of their publication record at the time the reward is granted. We have failed to study how other variables, such as personality attributes, have influenced the evaluation process. But, most important, it seems likely that the type of analysis we have conducted has failed to study the most significant basis for particularism: the positive and negative feelings of individuals toward other individuals and the location of scientists in networks of interpersonal social relations. If a scientist votes to grant a reward to a particular individual because that scientist likes the recipient or, conversely, votes to deny a reward to a particular individual because that scientist dislikes the applicant, this action would seem to be an example of particularism at work.[14] But if the sum of likes and dislikes is not correlated with any of the independent variables included in our studies of the stratification process (both statuses external to science and affiliations internal to science), particularism can be rampant but not show up in any of the statistical analyses. Because particularism does not, for example, result in older scientists being favored over younger ones or scientists from top-ranked departments being favored over those from lower-prestige departments does not mean that particularism is not operating as a result of network ties. I believe that this type of particularism is as widespread in science as it is in most other institutions. That statuses such as professional age and prestige rank of department show only small correlations with rewards, independent of role performance measures, suggests only that such variables are not strongly correlated with the strength and nature of a scientist's network ties. In the next chapter I will conduct a theoretical analysis of the conditions under which this type of particularism is most likely to come into play in science.

· · 8 · ·

Conceptualizing and Studying Particularism in Science

Before continuing our investigation of how the role of particularistic criteria in the evaluation of scientists and scientific knowledge can be studied, we must take a closer look at the concept of "universalism-particularism" in the light of new developments in the social studies of science. In order to study the importance of particularism, we must be able to classify an evaluation as being particularistic or universalistic. Consider the bases on which we develop our personal valences toward other scientists, and whether or not we should classify the use of such valences as particularism. If scientist A, in evaluating the work of scientist B, gave scientist B a negative evaluation because of scientist B's race, gender, religion, or other functionally irrelevant statuses, most of us would agree that this would be particularistic.[1] Most people would also probably agree that if scientist A gave scientist B a negative evaluation because of B's personality traits, this would be particularistic.[2] A serious problem arises, however, when we consider acting upon personal valences based upon cognitive criteria. Is it particularism, for example, to vote against hiring a new member of a department because you do not like the type of work that the scientist does? We might try to avoid this question by making a distinction between using technical criteria and substantive criteria in evaluating cognitive work. Most would probably agree that it is not particularistic to oppose a scientist whose work can be shown to be technically in "error." This criterion is only of limited use, however, because it is frequently difficult or impossible to reach agreement on such an evaluation. And what about opposing a scientist because his or her work follows a theoretical orientation different from your own, uses a methodology different from the one you prefer, or studies a problem which you believe to be trivial?

Consider a situation in which a sociology department is attempting to hire a new faculty member.[3] The current members of the department must evaluate a series of candidates. One candidate is a Marxist who conducts empirical analyses of elites, another does qualitative case studies of organizations in order to understand how structural barriers prevent women from achieving high positions, and a third uses sophisticated methodology to study the influence of the workplace on workers. In the ensuing discussion the following comments are made.

1. "The Marxist should not be hired, because his work is politically biased. He wants to show that a powerful 'capitalist class' exists in the United States and he picks and chooses evidence in order to support this conclusion, ignoring counterevidence."

2. "The sociologist who does qualitative case studies of organizations should not be hired, because his conclusions are not supported by any evidence. The study is conceptually and methodologically sloppy, making it impossible to determine the extent to which structural barriers or input factors influence the level attained by women employees."

3. "The sociologist studying the influences of the work context should not be hired, because his methods are overly complex. He cares more about the methods than about the substantive problem. So many methodological procedures intervene between the behavior of the people he studies and his conclusions that the validity of the latter is questionable."

For each of these statements there are some members of the department who agree and others who strongly disagree. Are these evaluations based upon particularistic or universalistic criteria? Is it particularism to believe that sociology should be "value free" or that sociology must be value oriented? Is it particularism to believe that concentration on quantitative methods has brought about a "displacement of goals" in sociology or that the only way to advance sociology is to increase the sophistication of quantitative methods? Is it particularism to believe that a particular theoretical orientation is wrong or a dead end? If so, how does one draw the line between evaluations based upon legitimate ("objective") cognitive criteria and those based upon illegitimate criteria?[4]

It is at this point that the critique made by social constructivists of work influenced by Merton's essay on the ethos of science becomes significant. In the early 1970s Merton's essay ([1942] 1957a) became the focus of criticism by sociologists influenced by relativistic philosophy and the new historiography of science (Barnes and Dolby, 1970; King, 1971; Whitley, 1972). These arguments were well summarized in Mulkay (1979). Merton's critics reject his theory of the ethos of science on several levels. First, they argue that it is not true empirically; the norms Merton describes do not actually govern the behavior of scientists. They base this conclusion on what they see as ample evidence that scientists do not in practice follow the norms.[5] Second, and most important, they argue that there are *no* "pre-established impersonal criteria consonant with observation" (Merton, [1943] 1957a, p. 553) which are used to evaluate scientific work. The evaluation of scientific work is inherently subjective, based upon the evaluating scientist's own intellectual interests and perspective. There frequently are no objective methods to determine which of two conflicting or competing interpretations is correct or matches nature. Third, given that scientific advance does not occur in the rational way depicted by the positivists, it is untrue that the norms of science are *necessary* in order to guarantee scientific advance.

This critique is used to attack the work on social stratification in science done by "Mertonian" sociologists of science. For example, Mulkay (1979) discusses *Social Stratification in Science* (J. R. Cole and S. Cole, 1973) as an example of the way in which positivism leads to misleading interpretations of how the stratification system works. Mulkay correctly points out that the monograph's argument is based upon the relatively high correlation between recognition and both the quality and quantity of scientific output and the low correlations between recognition and other variables. If the recognition is actually meted out on the basis of an objective evaluation in which preestablished and stable criteria are used, then the reward system is indeed based upon the norm of universalism. But if the notion that science is evaluated by this kind of objective criteria is questioned, then we must ask how the work of some people comes to be evaluated more favorably than that of others. What does it mean to suggest that some scientists' work is of "higher quality" than the work of others?

Are there perhaps systematic social differences in participants' ability to establish that their work is of high quality? Thus, although it is clear that female scientists' relatively low rank is associated with low quality of work (as recognised by other, predominantly male, scientists), we would no longer be forced to see this as the result of "objective" differences in the findings produced by male and female researchers (Cole and Cole, 1973). It has become possible to conceive that women and the members of other social categories in science are systematically prevented from (or favored in) establishing that their work is of high quality. By moving away from the standard view and the associated notion of "universalism," that is, by assuming that cognitive criteria in science may be flexible and their application to particular cases problematic, it becomes possible to investigate whether the social allocation of "quality," and thereby social rank, is affected by structural differences within the scientific community. (Mulkay, 1979, p. 26)

Mulkay raises a significant empirical question regarding the ability of various groups to succeed in having their work defined as of "high quality." No empirical work done by either critics or supporters of Merton's position on the ethos of science has demonstrated that the work of any particular social (as opposed to cognitive) group is differentially evaluated, using social or power considerations.[6] What we do know about women scientists is that their citation patterns tend to be similar to those of males; that is, women scientists cite the same work as do male scientists. It is thus very unlikely that the work of women scientists would be differentially evaluated because they are women. Zuckerman (1988) points out that because citations are made by rank-and-file scientists and are not the product of "deliberate collective decisions . . . it seems unlikely that the rank-and-file scientists who are responsible for most citations are motivated to reward the elite, beyond what they take to be their due" (pp. 529–530).

For contemporary American science no evidence exists, nor is it likely that any will be found, to support the claim that various *social* groups differ in their ability to have their work defined as of high quality. The Mulkay argument is more compelling when applied both to individuals and to *cognitive* groups. Some individuals in science are more effective than others in the use of interpersonal strategies that succeed in bringing favorable attention to their work. We must remember that at the research frontier there are only a small number of new contributions whose cognitive characteristics bring about almost immediate consensus and entry to the core. For the overwhelming majority of new work, either the community ignores it entirely or the level of underdetermination is so high that it is difficult to form a consensus on the value of the cognitive contribution. Because of this lack of consensus, all sorts of social processes influence the evaluation. In this circumstance the *promotion* of work may sometimes be a more important determinant of how it is evaluated than its cognitive characteristics. Scientists who are attuned to and adept at manipulating these noncognitive social processes will experience more career success than those who are not attuned or are less adept.

For example, science, like other institutions, depends upon the exchange of personal favors. Scientist A will be asked to write a letter of evaluation for scientist B who is being considered for promotion to a higher rank. Scientist A writes a highly favorable letter and sends a blind copy to scientist B. The following year scientist B is a referee of a grant proposal or journal article submitted by scientist A. He writes a favorable review. Is this really a "favor" and an example of particularism? It is quite possible that scientist A and B think very highly of each other's work. It is also possible that although they genuinely do think favorably of each other's work, their evaluations emphasize the positive for political purposes. Because of the importance of this type of "favor" exchange, it is increasingly difficult to persuade scientists to write critical appraisals of other scientists' work.[7] Because it is always possible that the identity of supposedly anonymous reviewers will become known, writing a negative evaluation can end up hurting the scientist who writes it as much as the scientist who is being evaluated. Given the low level of cognitive consensus, the more friends one has and the bigger one's network, the more likely it is that one will receive future rewards.

Not only some individuals but some cognitive groups will be more successful than others in promoting their work. In sociology, for example, there might be a difference in the ability of ethnomethodologists and social network analysts to have their work defined as of high quality. There are no objective criteria that allow sociologists to conclude that social network analysis is "more important" or of "higher quality" than ethnomethodology. Such a conclusion is a matter of subjective opinion. But should the use of cognitive opinions as the basis of personal valence be characterized as particularism?

Although as Laudan (1984) has pointed out (see Chapter 1), it is not true that there are no criteria which allow a choice between *any* two alternative scientific interpretations, it is usually true that for most work being considered at the research frontier there are generally no preestablished criteria by which conflicting scientific claims can be evaluated. There is extensive historical evidence to support this position, and I have presented quantitative data in Chapters 4 and 5 which indicate that at the research frontier science is marked by low levels of cognitive consensus. If there are frequently no preestablished criteria which allow for the unambiguous evaluation of truth claims made at the frontier, does this mean that science cannot be universalistic? If universalism requires the use of preestablished "objective" evaluative criteria, then the answer would have to be in the affirmative. The positivist position tends to assume that scientists make emotionally neutral unbiased evaluations of frontier knowledge. But as Mulkay (1979) has pointed out, it is

> not simply that complete neutrality is impossible, but that considerable commitment is necessary before even the simplest kind of observational work can begin . . . For, if as appears often to be the case, available criteria are unclear, or not easily applicable to particular instances, or different persons are using different criteria, it seems impossible for scientists to employ this principle [universalism] in practice; even though, once reasonable consensus has been achieved, scientists may be able to formulate *ex post facto* the criteria which they have finally agreed are appropriate to a particular body of knowledge. In other words, the sociological notion of "universalism" presupposes that technical criteria are generally available in science in such a way that firm, impersonal judgments can be made with respect to most knowledge-claims, and, thereby, with respect to the rewards and facilities which scientists deserve. (p. 65)

Once it is recognized that in science as it is developed at the research frontier there are generally few objective criteria that can be specified which will lead to agreement in evaluation, then the difficulty of empirically separating "objective" universalistic evaluation from "subjective" particularistic evaluation becomes evident.

Consider the difficulty in classifying the following case as representing universalistic or particularistic evaluation. Assume that a reviewer of a grant proposal in economics does not believe that Marxist economics can make any contribution to our understanding of economic problems. He is sent a proposal written by a well-known Marxist. He reads the proposal carefully and writes a series of criticisms of the design of the study, the methods to be used, and the theoretical utility of the results which will be attained. He does not, of course, say that the proposal should not be funded because the author is a Marxist. But is a universalistic evaluation of this proposal by this reviewer possible? A positive answer to this question would require two assumptions. First, one would have to assume that it is possible for evaluators to be objective about specific pieces of work which have cognitive characteristics (methodological, theoretical, or substantive) toward which they have negative personal valences. But as Mulkay and other relativist sociologists of science have pointed out, it may be impossible for a reviewer to be unbiased or to apply the same set of criteria to a work which he believes is fundamentally intellectually flawed as to one which he believes is sound. Second, one would have to assume that it is "illegitimate" to evaluate work based upon its general cognitive characteristics. This position assumes that the norm of universalism requires scientists to have an "open mind" and that all points of view *should* be tolerated, even those which one believes are completely fallacious. Would we expect a contemporary chemist to evaluate "objectively" a proposal from an alchemist? Would it be particularistic for most physicists to have a negative orientation toward a proposal to conduct further research on "cold fusion"? Making this sort of intellectual judgment on what is "right" or "wrong," promising or not promising, worth pursuing or a dead end is a necessary requirement of being part of the scientific community. The analysis of the proposal is thus particularistic, but only in the sense that virtually all evaluation of frontier science involves the application of subjective criteria which cannot be proved to be valid.

If we accept that most cognitive evaluations of work at the research frontier are subjective, it becomes difficult to classify evaluations made on the basis of cognitive likes and dislikes as particularism. If we do make such a classification, then we must recognize that science is thoroughly particularistic in how it evaluates others. In this case the distinction between particularism and universalism becomes meaningless and we must classify the reward system simply by the type of particularistic criteria that are being used. In essence this is what I am suggesting. This view of science begins by recognizing underdetermination as a variable. Very occasionally there are some problems whose level of underdetermination is low and whose cognitive aspects compel acceptance. That most contributions are either ignored as unimportant or have levels of underdetermination high enough so that cognitive aspects do not compel acceptance at least in the short run does not mean that cognitive aspects have *no* influence on the evaluation process. Even in the great majority of cases, where the evidence, models, and theories cannot compel acceptance and new contributions are met with disagreement, cognitive aspects will play some role in their evaluation. Only more detailed case studies will allow us to understand how cognitive aspects interact with the noncognitive aspects of the evaluation process.

That most scientific evaluation must be classified as subjective or "particularistic" does not tell us anything about the specific bases for particularistic judgments. The questions raised by "Mertonian" sociologists of science remain valid, but now, rather than asking whether science is particularistic or universalistic, we are asking what the social bases of particularistic evaluation in science are.

Research on universalism conducted by American sociologists of science, including my own previous research, is based upon the important assumption that the number of papers published and the number of citations to these papers are valid objective measures of the "quality" of a scientist's role performance. As I have pointed out, serious questions have been raised about this assumption by constructivist sociologists of science. But even if we recognize that indicators such as the number of citations to a scientist's work are influenced not only by the "quality" of the work but by social processes as well, we still want to know whether other statuses and affiliations of a scientist have an influence on how his or her work is evaluated, independent of the citation measures of scientific role performance. In Chapter 7 I argued that the research done on this topic by researchers such as Allison, Long, and McGinnis is valuable in helping us understand how the process of accumulative advantage works in science, but does not present convincing evidence that at the *institutional* level the evaluation of science is strongly influenced by either irrelevant nonscientific statuses such as gender or "irrelevant" scientific affiliations such as prestige of Ph.D. department or current department.

Scientific Networks and Particularism

The scientific community can be described as a series of social networks in which strong and weak ties (Granovetter, 1973) are based upon intellectual alliances, political alliances, institutional loyalties, and friendship (Collins, 1986; Latour, 1987). How a scientist is tied or not tied to those who end up doing the evaluation in a particular circumstance can influence the outcome. In this section I will describe those situations in which network ties are likely to play greater or lesser roles and the variables which influence the extent of importance of network ties. Because these network ties are based upon empirically overlapping cognitive and personal factors which frequently cannot be disentangled, we cannot assume that decision making resulting from location in a network of personal ties is strictly a result of "personal" as opposed to "cognitive" particularism. But whatever the bases of network ties, it is probable that such ties have a strong influence on the way in which scientific work is evaluated. If so they are part of an important social process influencing the competition among new contributions to become part of the communal knowledge base.

Four variables influence the extent to which a specific reward situation will be influenced by the particularism of network ties. First, to what extent do personal ties between the evaluators and the evaluated normally exist? The more prevalent such ties, the greater the chance that the ties will play an important role in the outcome of the evaluation. Second is the "observability" or "openness" of the evaluation process. When the deliberations are secret and evaluators are not required to justify their decisions, the chance that the particularism of network ties will operate is greater. The third variable is the scarcity of rewards; the more scarce a reward, the greater the likelihood that particularistic network ties will be used in decision making. Fourth, and perhaps most important, is the clarity of the criteria of evaluation. The more ambiguity in the criteria and the less consensus on them, the greater the chance for the operation of all types of particularism, including that of network ties. We may apply these four variables to a theoretical analysis of the extent to which six reward situations in science are likely to be influenced by network ties.

1. Admission to graduate school. When students apply for admission to graduate school, they are generally not part of the scientific networks to which the members of the admitting committees belong. In some cases there may be a weak tie between an applicant and a member of the admitting committee (for instance, the member may have relationships with some of the applicant's instructors), but in general there will be few network ties. In graduate school admissions the observability of the decision-making process is generally low, because each department employs its own review mechanisms and is not held accountable for its decisions. Today and even in the past the scarcity of places in graduate departments has been low relative to other rewards. There is some significant agreement on the criteria (college grades and Graduate Record Examination scores) that should be used in judging applicants.[8] With the exception of observability, the other three variables lead to a situation in which we would expect to find location in networks having a low level of influence on the outcome.

2. First job. We can expect recent Ph.D.s or postdoctoral fellows, when they seek their first job, to have more ties within the scientific network. Students will at least have ties with their sponsors and, through their sponsors, weak ties with sponsors' associates. Although there are usually no direct ties between candidates and department members, the anticipation of future ties may play a role in decision making. Department members will want to hire a new professor who they believe will be aligned with their cognitive and political interests within the department. The observability of the hiring process is greater than it is for the admission of graduate students, because one's colleagues and the administration must agree on appointments, but even so, the discussions that go on in recruiting committees are not highly observable. Today positions as assistant professor are scarce, ensuring that there are many more potentially qualified candidates than there are positions to fill. In addition, the criteria of evaluation are not clear. Young Ph.D.s have generally not had enough experience to be evaluated primarily on the basis of their publications. They therefore tend to be evaluated on their self-presentation and on the assessment of the hiring department of their scientific "intelligence." Such variables are hard to measure, and there will be differences of opinion among evaluators. In this situation we can expect to find that network ties will have a moderate level of influence on the outcome.

3. *Tenure.* By the time an assistant professor is ready to be evaluated for tenure, that person has either succeeded or failed in establishing meaningful ties with senior faculty in her department. These ties (positive or negative) will play a significant role in the judgments of the evaluators. The decision-making process on tenure is relatively open and easy to observe. The people who make the decision must be able to justify it to the candidate, her sponsors, and the university administration. The need for justification puts pressure on the evaluators to use "universalistic" criteria. One of these criteria is the quantity of publications, which is relatively easy to assess. More difficult to assess is the "quality" of publications, supposedly a more important criterion. Because there are usually no objective criteria enabling an assessment of quality, there will frequently be disagreement on this judgment and pressure to weight the quantity of publications more heavily. The scarcity of tenured positions for assistant professors is only moderate, as most universities (but not Harvard or Yale) promote many junior faculty to tenured associate professorships. The criteria in evaluating candidates for tenure are clearer than in evaluating candidates for new positions. Minimum "quantity" standards generally obtain; the quality of at least the journals in which the

candidate has published can be evaluated, and outside letters can be requested. The substance of the outside letters, of course, is based upon very subjective evaluations and may in turn be heavily influenced by the relationship between the candidate and the evaluators in social networks. In the case of tenure decisions we thus have one important variable pushing toward particularistic choice (the ties of the candidate to the evaluators) and several which push toward the application of more "universalistic" criteria (high levels of observability, only moderate scarcity, and relatively clear criteria of evaluation). The extent to which one of these forces will be greater than the other varies from department to department and even within department from case to case. But in general I would expect to find that network ties have only a moderate influence on the outcome of tenure decisions.

4. Awarding grants. There exist many different systems for awarding research grants. Here I will focus on that employed by the National Science Foundation. In the NSF process the potential for extensive personal ties between the applicant and the evaluator exists, although in many cases these ties will be minimal or nonexistent. Pfeffer, Salancik, and Leblebici (1976) have shown that in four social sciences the amount of funds received by a particular institution was correlated with whether or not that institution had representatives on the NSF panel which helped make the grant decisions. Although the mechanisms are not specified in their research, it seems clear that network ties with panel members can increase the probability of funding.

The observability of the evaluation process employed by the NSF is very high. The program director must provide rejected applicants with verbatim but anonymous copies of the reviews and be able to justify her decisions to her superiors. The resources are only moderately scarce. In some of the natural sciences about 50 percent of applicants are funded; in sociology this figure is generally closer to 25 percent. The clarity of criteria is not high, because the pervasive lack of consensus discussed in Chapter 4 allows different reviewers to define the same projects as meritorious or nonmeritorious. This combination of factors most likely leads to a situation in which network ties would have moderate to low levels of influence on outcomes.

5. Hiring senior professors. When senior professors are hired from outside the department, all the variables emphasize the importance of particularism. Well-known senior professors have significant ties within scientific networks. These ties can be very important, in some cases crucial, in the selection of candidates for senior positions. The observability of the evaluation process is relatively low. A department must present a "qualified" candidate. Clearly the majority of scientists will not be eligible for the position, but among those who are eligible there is usually no need for the evaluators to justify selecting one over another. In this case there is a great scarcity of positions and the clarity of criteria is low. The candidate must have made "significant" intellectual contributions to the discipline. But given the low level of cognitive consensus in science it is almost impossible to find candidates whose work everyone will agree is "truly" outstanding. There are a few such cases, but these people are in such demand that in most hiring situations they are rarely the candidates being considered.

This situation illustrates the potential connection between the scarcity of the reward and the clarity of the evaluation criteria. Imagine that you are a professor considering a series of sociologists for two different rewards: election to the Sociological Research Association (SRA), an honorary organization limited to 150 active members, and appointment as full professor in your department. There are probably many individuals who you would grant "deserve" to be elected to the SRA but whom you would not want to be full professors in your department. That many can be elected to the SRA makes it easier to reach a working consensus. That only one person can be appointed to your department increases the difficulty of reaching consensus on whether a particular individual does or does not meet the criteria. In the latter case more fine distinctions will be made, which increase the chance that the decision will be influenced by network ties.[9]

6. *Prestigious awards.* Like the hiring of top-level senior professors, the granting of prestigious awards is a situation in which generally all the variables lead to a high level of expected particularism. By the time a scientist is "ready" to receive a prestigious award she generally has extensive ties with scientific networks, and these ties can help or hinder receipt of the award. The process leading to the award is usually not observable; sometimes deliberations are carried out in secrecy. Prestigious awards are also extremely scarce; there are many more qualified scientists than available rewards. Once again the clarity of the criteria are low. There is room for endless disagreement over whether a particular candidate's work is "truly" outstanding. We would expect network ties to be very important in the distribution of these rewards.

The Two-Cut Process

The hiring of top-level senior professors and the granting of prestigious awards such as the Nobel Prize or MacArthur Fellowships brings to light a situation, not infrequently found in science, which might be called the "two-cut process." Sometimes in science there is a two-stage process of decision making: in the first, a pool of eligibles is defined, and in the second, one or more candidates are selected from among the pool of eligibles.

Consider, for example, the awarding of a Nobel Prize or admission to the National Academy of Sciences. It seems on the surface strange to classify the distribution of such rewards as being heavily influenced by one's location in scientific networks. Virtually everyone who receives these rewards is highly deserving. The problem lies not in the selection of candidates who do not qualify (this would usually be impossible even if the evaluators wanted to do so) but, rather, in the selection from among the *many* candidates who do qualify. In reward situations of this type the decision making tends to be done in two steps. First, a group of candidates is nominated. It is unusual for an unqualified candidate to be nominated. Such a nomination would reflect poorly on the nominator; it is also known that the ultimate prize winner must be presented to the world and must have impeccable qualifications. Thus, although not all qualified candidates are nominated, it is likely that most of the nominated candidates are qualified. In the second stage a selection must be made from this group of candidates. When only one or a very few must be selected from a much larger group of qualified applicants it may be impossible to select one applicant or aspirant over another on the basis of a "universalistic" principle.

Zuckerman, in her study of Nobel laureates in the United States (1977a), explicitly discusses this duality in the selection of Nobelists:

> The question of whether universalism or particularism governs the allocation of the prizes is badly put. Each may be applied or both, first one, then the other. The model of decision making for the Nobel prize may be thought of in terms of successive phases of selection from smaller and smaller pools of candidates. In the first phase, significance of scientific contribution should take precedence in the sorting process since laureates, on the whole, are generally considered to have made major advances in their fields. Only rarely, however, does the first cut generate a small number of decisively superior contributors who stand out above the rest . . . But generally, the first cut produces a cadre of candidates

who on a first approximation seem pretty much on a par. Since some additional bases for selection are required if a choice is to be made, secondary criteria are called into play: some of these remain functionally relevant to the advancement of scientific knowledge but others are particularistic. (pp. 48–49)

These particularistic criteria have included "the candidate's nationality, politics, or even their affability, rectitude, or the tidiness of their domestic lives" (p. 49). But the ultimate selection of one candidate rather than another may also be based upon particularly strong preferences, likes and dislikes, and interests of members of the selection committees. The potential consequences of particularistic final judgments on the fads, fashions, and research choices and styles that result from the importance attached to the Nobel Prize should not be overlooked. Once selected, a Nobel Prize winner's influence on the future development of his or her field can be great. That effect can multiply well beyond what it would have been without the award. Thus the influence of particularistic network ties will make a great difference for the individuals selected as well as for the cognitive development of a research area.

Problems in Empirically Studying the Bases of Particularism

In analyzing the difficulty in determining whether to classify a particular evaluation as being based upon particularistic or universalistic criteria, I have concluded that most evaluation in science must be classified as particularistic, but that at least analytically it is possible to distinguish several different bases for particularistic judgments. These bases would include both irrelevant nonscientific statuses and scientific affiliations of the scientist being evaluated as well as personal valences based upon a wide array of determinants, which range from cognitive evaluations of the scientist's work to the scientist's personality, political views, or institutional connections with the evaluator.

It is possible to study the extent to which location in scientific networks can influence the evaluation of a scientist's contribtuions. We might be able to show, for example, that the size of a scientist's friendship network has an influence on the evaluation of her work independent of its "quality." This task would be difficult, though not impossible, because we would need a measure of the "quality" of the work which was based upon a "blind" evaluation, similar to those I used in the NSF peer review experiment (see Chapter 6). Let us assume that we could establish that among a group of scientists whose work was of the same "quality," those with larger friendship networks were more likely to receive favorable evaluations and greater recognition. We would then want to study the mechanisms through which involvement in the friendship network influenced the evaluation or the bases of the "particularistic" evaluation. Unfortunately, at this point we are likely to run into a serious problem—the empirical intermeshing of cognitive and noncognitive influences on personal valence. Consider two scientists who get to know each other because each finds the work of the other useful. They become good personal friends. The personal friendship would, of course, influence their evaluation of the other. It would then be empirically impossible to separate the cognitive from the noncognitive sources of the personal valence.

To illustrate the difficulty that would be encountered in attempting empirically to separate cognitive from noncognitive influences on the evaluation of scientific work, let us consider the findings on the NSF peer review system reported in Chapter 4. From the statistical analysis of the data and detailed qualitative evaluation of the content of more than 1,000 reviews, we concluded that differences in the way in which reviewers evaluated the same proposal were generally due to genuine differences of intellectual opinion. But it is also possible that differences in the evaluation of the same proposal could result from differing personal valences of the various reviewers toward the principal investigator. One reviewer of proposal X may have a positive personal valence toward its principal investigator whereas another reviewer of proposal X may have a negative personal valence toward its principal investigator. This is undoubtedly true in some cases of investigator-reviewer pairs, but the extent to which such personal valences may be based upon cognitive concerns or noncognitive concerns is not known and in practice would be hard to determine.

The following example is drawn from a proposal in a social science discipline but is not significantly different from many similar sets of reviews of proposals in the natural sciences. There were ten reviews of the initial proposal: one "excellent-plus," one "excellent," three "very good," one between "very good" and "good," three "good," and one "poor." Consider four positive comments:

Reviewer 1: I give this proposal an "Excellent-plus" rating. The measurements to be developed will, if successful, provide extremely useful tools for the assessment of —————— in several dimensions, and I am fully confident that the effort will succeed. I do not think it an exaggeration to suggest that such new measures will afford advances in understanding science comparable to the advances in biology made possible by the substitution of the electron microscope for the optical microscope . . . This reviewer is acquainted [with] both of the [principal investigators] and has admired their work over the past ten years. They have been at the forefront of ——— and this proposal testifies to their clear sense of what the field needs now in order to advance . . . The proposal has my complete and enthusiastic endorsement. (rating: Excellent plus)

Reviewer 2: This is a well-thought through proposal which addresses itself to issues of great significance . . . On the whole the proposal promises to look at very important problems—problems of great significance to the academic community and to policy makers. (rating: Very Good)

Reviewer 3: This is an extremely competent and carefully designed project which should yield substantial scientific payoff. It is in the tradition of much good work in recent years in the area of ———. I basically expect that this project will produce the best measures yet available of the ———. Clearly this type of measurement is not and should not be the end of the story and additional work is needed to interpret the evidence of these measures . . . But this is a very sound first step. The project passes with flying colors on criteria IV and V. (rating: Very Good)

Reviewer 4: I am very impressed with the design and objectives of this proposal. It seems to me of the highest quality and importance and is a major desideratum for further advances in ———. As to the applicants, their recent work on ——— has been, I suppose, the most impressive of anybody in the world . . . I give the entire proposal full marks with absolutely no reservations. (rating: Excellent)

The enthusiasm displayed here is not matched, however, by several of the other reviewers. Consider the judgment of two reviewers who did not see merit in the proposal:

Reviewer 5: This proposal has to be evaluated in terms of the very highest scientific standards . . . The proposed project would build an excessively elaborate, expensive, and questionable statistical analysis on a foundation of derivative empirical data that has not yet been shown to be extensive, and consistent enough to support significant findings, or have the resilience to provide reliable indicators in the future . . . As a scientific inquiry, the proposed project has fundamental weaknesses that are not clearly acknowledged . . . As it stands, this proposal is not methodologically sound, despite its preoccupation with ways to anticipate and correct for potential errors resulting from ———. The budget is very large in proportion to the new knowledge this project would produce . . .

[This summarizes two full, single-spaced pages of comments.] (rating: Poor)

Reviewer 6: This proposal is long and detailed, indeed even prolix. It offers strong evidence of the industry, determination and tenacity of the principal investigators. [They] are well-respected students of ———. However, I am far from persuaded that it would be wise to commit ——— of NSF funds to the project they outline . . . This proposal veers uncertainly between grandiose aims . . . and claims . . . and some rather modest specifics . . . Perhaps a modest grant for a pilot project would be in order. (rating: Good)

After more than a year of deliberation and another round of reviews, the proposal was rejected for funding. Was this decision based upon cognitive or noncognitive particularistic criteria? We see from the reviews that the principal investigators were known either personally or through their work to virtually all of the reviewers. One reviewer specifically mentions that he was acquainted with the principal investigators; another refers to them as well-respected students of a leading scientist in the discipline. Another says that their recent work on the topic of the proposal has been among "the most impressive of anybody in the world." We may assume that not only did all or most of these ten reviewers know the principal investigators but most of them had personal valences toward them; that is, they either liked or disliked them. We can go no further with the analysis. We cannot determine from these data the extent to which the personal valences were based upon "genuine" intellectual criteria or nonrelevant personal or political animosities in the field.[10]

The lack of cognitive consensus in science offers a potential "shield" for the utilization of noncognitive particularistic criteria. Because disagreement on cognitive issues is widespread in science, it is possible for any evaluator with a negative or a positive personal valence toward the evaluated scientist, no matter what its basis, to shroud this valence in a cloak of cognitive "rationalizations." It is, of course, possible that such personal valences could be based upon what the evaluator interprets to be legitimate cognitive criteria. Empirically "legitimate" cognitive criteria and "illegitimate" personal criteria may be inseparable. The peer review study did not offer the opportunity to determine the extent to which the observed disagreements were based upon "legitimate" or "illegitimate" criteria. The data did show that there was no *systematic* bias, that is, that the overall reversal rate was no different from a chance distribution which could be predicted from knowledge of the extent of disagreement (see Chapter 4 and the Appendix).

The various scientific fields probably have norms which specify the boundaries of what is "legitimate" or "illegitimate" evaluation. It is possible that the style of the evaluation may be more important than its content. Because scientists believe that science is "supposed" to be universalistic (not necessarily that it "is" universalistic), evaluations which are couched in the most objective-sounding language are likely to be more effective in convincing others. Strong language and *ad hominem* remarks will be less credible because they give the appearance of particularistic bias. In face-to-face situations the tone of voice and facial expression of the evaluator may be just as significant as the content of her evaluation in creating the appearance of objectivity. Studies of the style in which evaluations are made and their effectiveness would be useful in developing a deeper understanding of the importance of social factors in the formation of cognitive consensus in science.

Authority and Evaluation

I have been analyzing the social processes which influence the evaluation of a particular scientist on a particular reward decision—for example, whether a particular scientist should be granted tenure in department X or whether a particular proposal should be funded by the NSF. Micro-level evaluations are very important to the individual scientists whom they affect. They also influence the broader macro-level evaluation process. I have emphasized the importance of *communal* knowledge, knowledge which is accepted as true by the scientific community at a given point in time. Now I address the crucial question of how, through the operation of intellectual authority, the thousands of micro-level evaluations affect the emergence of consensus at the community level.

I believe that the social processes through which work at the research frontier is evaluated are similar in the various scientific fields. How does each scientist decide what he or she should accept as important and valid or reject as uninteresting or wrong? A positivist answer to this question would focus on how each individual scientist evaluates for herself the contributions of others. According to this view, the individual, using commonly held criteria, determines what is good and what is not. Reputations are the sum of these individual evaluations.

All scientists do have some opinions which are based upon their own direct and independent appraisal of scientific work. But to a large extent our opinions of what good science is and who has done good work are based on judgments made by other people or by the evaluation system in our field. Think of sending scientists a questionnaire (similar to that described in Chapter 5) asking them to evaluate the quality of work of a sample of their colleagues. Then think of interviewing each respondent to find out how much detailed knowledge he had of each scientist he evaluated on the questionnaire. We would undoubtedly find that most scientists have an opinion of the work of many others which is based upon minimal direct knowledge and experience. The judgments of the scientific community are frequently accepted without the benefit of personal confirmation. The ways in which these authority practices work require detailed investigations. Here I give a few illustrative examples.

First, consider even the evaluations which are made by individual scientists. Rather than these evaluations being based upon some objective set of rules which all scientists learn, they are based upon subjective and sometimes idiosyncratic criteria which most scientists internalize in the process of receiving their graduate and postgraduate training. Most scientists have been educated at a relatively small number of prestigious graduate departments, where they studied with eminent scientists. To some extent the views of these teachers influence the standards adopted by their students and the subsequent evaluations their students make. Thus the intellectual tastes of one charismatic eminent scientist can directly influence the evaluative criteria used by hundreds of her students and indirectly those used by the students of these students.

As I pointed out in Chapter 7, many researchers in the sociology of science have found evidence of accumulative advantage. Studies I have conducted on scientific reward systems indicate that evaluations made by scientists of other scientists are influenced by the rated scientist's departmental rank independent of the "quality" and quantity of her publications (J. R. Cole and S. Cole, 1973). Thus if we consider two groups of scientists who have published work of approximately equal quality, but one group is located in prestigious departments and the other in less prestigious departments, on average the work of the former will be more favorably evaluated. Why does this happen; is it simply because we look up more to those at prestigious institutions? Although this may be true to some extent, I suggest that the primary reason for accumulative advantage is the operation of intellectual authority. When colleagues say that of two scientists with equal productivity and equal citations, the one at the more prestigious department is better, they are not simply giving that scientist extra points for being at a prestigious department. What is more likely is that they believe that the work of the scientist at the prestigious department *is* "better" than that of the one at the less prestigious department, because they accept or internalize the judgments of important authorities. Authorities can be both individuals and institutions. When prestigious institutions, whether they be departments or agencies or foundations, bestow recognition on a scientist, their decision affects the way in which others perceive that scientist's work. If a prestigious department thinks that X is good enough to hire, she must be better than Y who works at a low-prestige department. It is in part because our evaluations are based on judgments made by others that the process of accumulative advantage exists. A favorable evaluation by authorities at Time 1 will influence the evaluation made by others at Time 2 independent of the "quality" of role performance in the intervening period. Data to support this view were presented in J. R. Cole and S. Cole (1973, pp. 116–122). There we showed that even when the number of citations was controlled, the rank of a scientist's department had an independent effect on visibility. This effect was substantially reduced, however, when we controlled for the "perceived quality" (a variable derived from questionnaire assessments of the scientists in the sample—not citations). Thus of two scientists who have the same number of citations (an "objective" indicator of role performance), the one located at the prestigious department will be *perceived* as having done higher-quality work (a subjective assessment of role performance).

Many more examples of the ways in which our scientific judgments are influenced by evaluations made by authorities can be mentioned. Scientists read papers in journals only after they have been evaluated by others. Because prestigious journals act as organizational authorities, scientists receive more credit from their peers for publications in prestigious journals. Scientists who have received grants, fellowships, awards, and membership in prestigious organizations are more highly thought of. Again, this is not simply because the community gives prestige to the recipients of awards, although this is to some extent true. Rather, the reward granted by an authoritative institution has both a direct and an indirect effect on the way in which the recipient is evaluated. The direct effect is that the recipient gets more prestige because people respect the reward. The indirect effect is that the quality of the recipient's work is evaluated more highly *because* of the reward. Thus an evaluator may believe that being elected to the National Academy of Sciences, for example, is as much a matter of politics as of science, but when the evaluator reads the work of a newly elected NAS member the evaluator cannot avoid having his or her judgment influenced by the background knowledge that the author is now an NAS member. There is no empirical way to separate these direct and indirect effects, but both are the result of the operation of intellectual authority.

The dependence upon the opinions of others is also illustrated in the process of hiring new colleagues. Rarely do all the members of the department sit down and thoroughly read the scientific work of the candidates. Usually a recruiting or search committee is set up, and in most cases only a portion of its members read a small portion of the work of each candidate. Later on, when the recommendations of the recruiting committee are discussed at a department meeting, faculty members make favorable and unfavorable comments about the various candidates. These comments are only infrequently based upon a detailed analysis of the candidates' cognitive contributions. They frequently consist of only evaluative statements about the work or statements about the candidates' personality or reputation. Those colleagues who have not made an independent evaluation of the candidate's work will depend more upon the opinions of some colleagues than others. Within each department there are generally some individuals who wield considerably more intellectual authority than others. Intellectual authority should not be confused with political authority, though both can affect appointments. The former affects judgments of the quality of scientific contributions; the latter influences judgments based upon other criteria such as departmental politics.

In the process of evaluating scientific work, the opinions of some people count more than the opinions of others. Generally the most important gatekeepers or evaluators are eminent scientists or "stars" who exercise legitimated intellectual authority. It is generally the stars in a field who have the greatest influence in determining who will be given appointments in prestigious departments and who will receive honorific awards and be admitted to honorific societies. One of the primary mechanisms through which the scientific community attempts to establish consensus is the practice of vesting authority in elites. When a new idea is proposed, it must be evaluated. Should the new idea become part of the consensus or should it be discarded? In the process of evaluation, some opinions count more than others. By their acts as gatekeepers and evaluators, those who are most eminent determine what work is considered good and what work unimportant. The mechanisms through which this type of authority influences evaluation require detailed empirical analysis. And, of course, the question of why the authorities select some work for promotion rather than other work also remains.

Because of the need for legitimated intellectual leaders, I speculate that the stratification systems of the various sciences are structurally organized in roughly the same way, regardless of their state of cognitive development or the degree to which stars have emerged naturally. A field which does not have "natural" stars will have to create them. The created stars will be those producing the "best" examples of the most fashionable style of work in the field at the time, even though their "absolute" contributions may not be as great as those of the stars who have emerged naturally in other fields or the stars in their own field who preceded them.

Conclusion: The Consequences of Particularism

In discussing the consequences of particularism for science I want to make an analytic distinction between particularism based upon cognitive evaluations and particularism based upon noncognitive influences on personal valences and network ties. Cognitive particularism is an evaluation which is based upon subjective as opposed to objective criteria. Because most evaluation of frontier science is based upon subjective criteria, most evaluation is particularistic in this sense. This type of particularism is an inherent part of science. Because nature does not speak to us in a clear voice, the attempts to read nature must inherently involve subjective elements. To say that scientists *should* be objective in their evaluation of others' work, as some normative philosophers of science have done in the past, is probably not fruitful. Historians and philosophers of science have made a fairly convincing case for the argument that the progress of science, rather than being hampered, is sometimes enhanced by subjective cognitive evaluation. In their view, if scientists actually followed the rational ways of doing science put forward by traditional positivist philosophers, it is unlikely that progress could be made.

Most observers of contemporary science agree that consensus is a necessary, if not a sufficient, condition for scientific advance. This thesis is clearly articulated in the work of Polanyi (1958), Kuhn ([1962] 1970), Ziman (1968), and Lakatos (1970). Kuhn suggests that science can make progress only after consensus on a paradigm emerges. Without that consensus, scientists cannot build on a corpus of completed work that is accepted as the given state of knowledge. Even though the rational behavior advocated by traditional philosophers would require a scientist to abandon a theory in the light of negative evidence, Kuhn argues that it is appropriate, because consensus on the paradigm is necessary, that scientists ignore anomalies until a new paradigm can be adopted. The maintenance of consensus requires a subjective or particularistic "irrational" attachment to the existing paradigm.

Polanyi (1958), in his book *Personal Knowledge,* gives many examples of scientists who continued to maintain their adherence to theoretical positions in the face of significant negative evidence or anomalies. But whereas some might find such behavior opprobrious, Polanyi argues that because consensus is necessary for scientific progress, scientists *must* frequently ignore anomalies in order to maintain consensus and, most important, that this behavior facilitates rather than impedes scientific progress: "It is the normal practice of scientists to ignore evidence which appears incompatible with the accepted system of scientific knowledge, in the hope that it will eventually prove false or irrelevant. The *wise* neglect of such evidence prevents scientific laboratories from being plunged forever into a turmoil of incoherent and futile efforts to verify false allegations" (p. 138, italics added).

Polanyi (1963) relates a story from his own experience. As a young chemist he proposed a theory of adsorption which was contrary to the theories currently accepted. At the time his theory was put forth, it was seen by other chemists as "wrong" and was essentially ignored. Twenty-five years later it turned out that Polanyi's theory of adsorption was correct. Polanyi himself, however, argues strongly that it was right for the scientific community to reject his idea at the time he first proposed it. Most new and contradictory ideas prove to be of little value. If scientists were too willing to accept every unorthodox theory, method, or technique, the established consensus would be destroyed, and the intellectual structure of science would become chaotic. Scientists would be faced with a multitude of conflicting and unorganized theories and would lack research guidelines and standards. So important is the maintenance of consensus, argues Polanyi, that it is better to reject an occasional idea which turns out to be correct than to be too open to all new ideas at the expense of preserving consensus.

Whereas scientists frequently maintain their adherence to paradigms in the face of negative evidence, in revolutionary situations they sometimes adopt a new paradigm on what must essentially be considered subjective cognitive criteria. The philosopher Paul Feyerabend, for example, argues that if empirical evidence had determined which theory on the system of planetary motion would be accepted, scientists would not have abandoned the geocentric theory when they did. In his analysis, the Copernican theory had more empirical problems at the time of its adoption than did the Ptolemaic theory. But the Copernican theory, "being in harmony with still further inadequate theories . . . gained strength, and was retained, the refutations being made ineffective by *ad hoc* hypotheses and clever techniques of persuasion" (quoted in Brush, 1974, p. 1166).

Feyerabend argues that faith in the Copernican theory in the face of negative evidence was necessary if new and revolutionary ideas were to emerge:

> It is clear that allegiance to the new ideas will . . . be brought about by means other than arguments. It will . . . be brought about *by irrational means* such as propaganda, emotion, *ad hoc* hypotheses, and appeal to prejudices of all kinds. We need these "irrational means" in order to uphold what is nothing but a blind faith . . . an unfinished and absurd hypothesis . . . What our historical examples seem to show is this: there are situations when our most liberal judgments . . . would have eliminated an idea or a point of view which we regard today as essential for science . . . The ideas survived and they can *now* be said to be in agreement with reason. They survived because prejudice, passion, conceit, errors, sheer pigheadedness, in short all the errors which characterize the context of discovery, *opposed* the dictates of reason . . . *Copernicanism and other "rational" views exist today only because reason was overruled at some time in their past* . . . Hence it is advisable to let one's inclinations go against reason in any circumstances, for science may profit from it. (quoted in Brannigan, 1981, p. 7)

In short, the new historians and philosophers argue not only that science is conducted in ways which could be described as particularistic and irrational but that acting this way is at least occasionally necessary for scientific advance.

What consequences for science does particularism based upon the noncognitive characteristics (such as gender, rank of doctoral department, personality, and political views) of those being evaluated have? (In asking this question we must keep in mind that the distinction between the two types of particularism is an analytic one which in practice will be empirically intertwined.) This type of particularism in science, as in any institution, can clearly have negative consequences for the individual "victims" and positive consequences for the individual "beneficiaries." But it is not always true that social patterns which have negative consequences for individuals also have negative consequences for social institutions. I believe that particularism based upon network ties and noncognitive characteristics has some serious negative consequences for science, but that these negative consequences at least in the United States are sharply reduced by the social organization of contemporary science.

Particularism has negative consequences because it is demoralizing for those working in science to believe that there will not be a close connection between scientific role performance and scientific recognition. To the extent that this ideal is breached it is possible for even those scientists driven by the "sacred spark" to have that spark dimmed. Why put in the hard labor to make difficult scientific advances when others less deserving are elevated to positions of fame and fortune? This question will undoubtedly be raised when the particularism of network ties becomes more visible and a dominant mode of evaluation. I would emphasize that this type of particularism has very negative consequences for the motivation and position of some scientists, and that what is said below should not be read as an apologia for it.

First, I should point out that although science must be considered to be predominantly particularistic in the way in which work is evaluated, the evidence suggests that noncognitive bases of evaluation such as nonscientific statuses (for example, gender) and irrelevant scientific affiliations (for example, rank of doctoral department) play a relatively small role in the way in which work is evaluated.[11] When we considered the importance of network ties based upon cognitively and noncognitively determined personal valences, I speculated that in some reward situations (for example, admission of graduate students or hiring new assistant professors) these would play a limited or moderate role. Two of the reward situations I analyzed, appointment to high-level senior positions and the distribution of scarce prestigious awards, are particularistic in a special sense. The first cut or first stage of evaluation for these awards is generally done on the basis of cognitive considerations; thus, unlike what occurs in many other institutions, the particularism of network ties in many situations in science plays a strong role only among the selection of *qualified* candidates.

It is possible to distinguish between a "universalistic" bureaucratic system of evaluation and a particularistic system. In the former an attempt is made to keep the rewards open to all qualified candidates and to evaluate all candidates using the same criteria. The system must have the appearance of being "fair." In a particularistic system of evaluation no attempt is made to open the system to all qualified candidates, and more idiosyncratic criteria are used in evaluation. This latter type of reward system is not necessarily harmful for the advance of knowledge. There is a difference between whether a reward system is "universalistic" and whether a reward system is "effective." A universalistic reward system is one in which the same set of functionally relevant criteria are used for evaluating everyone. An efficient reward system is one which gives resources to those who can best use them, even if not everyone is treated equally. The method of peer review used by the NSF is essentially a universalistic *system,* even though an evaluation made by an individual referee may involve particularistic criteria. This universalistic way of distributing grant money can be compared with more particularistic methods used by some private foundations and by governmental agencies such as the Office of Naval Research.

These latter systems depend less upon ratings received by a set of independent referees and more on the personal judgment of a program director. A program director may have a set of researchers who she believes are doing important work and who are funded on a regular basis. That the program director thinks a researcher's work is significant is all that is generally necessary for the researcher to get funding. Frequently grants are made on the basis of short letters rather than formal proposals. This system is definitely less "universalistic," because it is more difficult for someone who is unknown to the program director to "break in" and receive funds. The money is not equally available to all applicants. But in those cases where the program directors have good scientific judgment or taste it is quite possible that the particularistic system will be more efficient in advancing knowledge than the universalistic one.

There are three reasons for this. First, a particularistic system allows the research scientists to spend substantially less time in writing proposals. The chance elements involved in getting a grant means that the proposals of even the most eminent and brilliant scientists are turned down. In order to maintain funding it is necessary to submit many proposals. Some active scientists have complained vociferously over the last fifteen years that they must spend at least a quarter of their time writing proposals. The research on the NSF peer review system suggested that there frequently was little connection between the proposal and the actual research conducted. The proposal is seen by many as an unproductive essay writing contest.

Second, universalistic systems like that used by the NSF tend to treat the opinions of all reviewers equally. Program directors *say* that they do not weight each review equally in making their funding decisions. But the data show such a strong correlation between the mean rating received from the mail reviewers and the ultimate funding decision that we must conclude that for most purposes reviews are treated as being of equal significance. Thus the opinions of a reviewer who may have only scanned a proposal and misinterpreted some parts of it will count as much as those of a reviewer who has spent several days carefully studying the proposal. In a more particularistic review system the program director has more leeway in discounting some reviews and emphasizing others.

Third, it is also possible that a system such as that used by the NSF will penalize creative work at the research frontier. New ideas in science frequently conflict with existing ideas. Given the high levels of cognitive disagreement at the research frontier it is likely that innovative research proposals will receive some negative reviews. These negative reviews frequently lead to the rejection of innovative but controversial proposals. Such proposals may have a higher chance of being funded in a particularistic reward system.

Because of these problems with universalistic review systems we are led to the conclusion that a particularistic evaluation system in which a program director gives grants to people whom she knows and whose work she respects may be beneficial and not harmful for science. This type of system could also, of course, be harmful, depending upon whether the recipients of the grants are doing "important" or "unimportant" work. "Universalism" and "fairness" are "good" ways to distribute rewards because they fit our general value system, not because they have been empirically demonstrated to be the most efficient means for advancing scientific knowledge.

Another factor militating against the negative effects of particularism is the "functional equivalence" of many scientists. Examinations of the distribution of scientific productivity (J. R. Cole and S. Cole, 1972; Price, 1963) illustrate the highly skewed nature of contribution to science. Only a small number of scientists make scientific contributions which truly influence the development of knowledge. If one scientist is rewarded and elevated on particularistic grounds over another scientist, probabilistically the chance that this act will have an influence on the future of science is very small.

Consider the NSF peer review study, in which we found that if a second independent panel of reviewers were selected there would be a reversal rate of between one quarter and one third (see Chapter 4). For the 25 percent who, owing to chance, either are or are not funded, this decision is a vital one. But does this inevitable arbitrary element in the operation of universalistic and bureaucratic reward systems have any influence on the advance of science? Given the high level of self-selection which influences who applies to the NSF, we found that the funding decision was generally made from among a group of roughly equally qualified applicants. Therefore funding decisions which may be based upon individual preferences or "biases" of reviewers will not produce a significantly different aggregate level of quality of funded proposals. Evidence to support this claim can be found in the very low correlations observed between the prior track record of applicants and the funding decision, and in studies of the citations to work produced from funded and unfunded proposals, which show rather low correlations between funding and subsequent impact of published work (J. R. Cole and S. Cole, 1985; Carter, 1974).

Finally, there is one structural attribute of contemporary science, at least in the United States, which vitiates the potentially pernicious consequences of particularism. In the United States science is a highly decentralized institution.[12] Crucial evaluations, rather than taking place at one or just a few points, occur in literally dozens if not hundreds of points. In a major scientific field such as physics there are well over 100 university departments with doctoral programs. If a qualified scientist is denied tenure at one department on particularistic grounds, that scientist still may be hired by one of the other institutions.

In the United States there are multiple sources of research funding. Although the number of sources varies by specialty, with fewer sources being available in a field such as experimental high-energy physics than in a field such as sociology, in most fields scientists have the opportunity to submit their grant proposals to many potential funders. Similarly there are many journals to which articles can be submitted and multiple sources of awards and fellowships. Because evaluation is done at hundreds of points and because these evaluations are only loosely connected, it is highly unlikely that an individual doing work which any significant portion of his or her colleagues believe is important will be denied all recognition. Although it is easy to think of cases in which an individual suffered at the hands of particularism, it is difficult to think of more than a few individuals who have made even modest contributions to knowledge and who have been blocked out of all rewards owing to the operation of particularism. Given the highly skewed distribution of contributions to knowledge, there are, even in the current tight job market, substantially more senior professorships at prestigious universities than there are genuine "stars" to fill those professorships. Decentralized evaluation systems would seem to reduce the consequences of particularism. If this is true, it might mean that societies that have more highly centralized evaluation systems in science would suffer more negative effects, unless other mechanisms reduced the operation of particularism at individual levels. This proposition merits further investigation.

· · 9 · ·

Social Influences on the Rate
of Scientific Advance

In examining the growth of scientific knowledge, I have concentrated
on how social variables might influence the substance of scientific
discoveries.[1] But as we saw in Chapter 1, sociologists also study the
growth of knowledge by determining how social variables influence
the foci of scientific attention and the rate of scientific advance. In
this chapter I am concerned with the latter topic. Whether science is
socially constructed or discovered, there are significant variations in
the amount of science produced by different nations during different
historical periods. Here I am interested not in how social processes
influence the content of what comes to be accepted as scientific knowledge but in those variables which determine the *amount* of scientific
knowledge produced in a given society at a given period of time. I
develop a simple but general theory of the determinants of advance
and relate the results of a study which helped formulate the theory.

There is considerable variation in scientific advance. Although in
general science has grown at an exponential rate over the last few
centuries, the extent to which science prospers in different countries
at different times is a variable. Only a handful of countries have made
significant contributions to the current body of scientific knowledge.
Besides the United States these include Canada, France, Germany,
Great Britain, Italy, Japan, and the former USSR. Most underdeveloped countries have contributed very little to scientific knowledge.

In two important papers Price (1986; Price and Gursey, 1986) has
used the nation as the unit of analysis and counted the number of
authors listed in the *International Directory of Research and Development
Scientists*. This directory, published by the Institute for Scientific In-

formation, lists by country the name of each principal author of papers listed in *Current Contents* (a journal listing all the papers published in more than 1,000 scientific periodicals). Price's tally provides
a measure of the number of scientists from each country who published in international journals in a given year. Price (1986) then
correlates the number of scientists with the size of the country's gross
national product. He concludes: "The share each country has of the
world's scientific literature by this reckoning turns out to be very
close—almost always within a factor of 2—to that country's share of
the world's wealth (measured most conveniently in terms of GNP).
The share [of scientific papers] is very different from the share of
the world's population and is related significantly more closely to the
share of wealth than to the nation's expenditure on higher education"
(p. 142). Price and Gursey (1986) show a high correlation between the
number of scientific papers produced in a country and the amount
of electrical power consumed.[2] Although there is a high correlation
between GNP and a country's scientific output, the correlation is not
perfect.[3] Countries such as the United Kingdom, Switzerland, and
especially Israel have a much larger portion of the world's scientists
than could be predicted from their GNPs. The United States has a
somewhat larger portion of scientists and the former Soviet Union a
significantly smaller portion of scientists than could be predicted from
their GNPs.[4]

Price's analysis does not consider variations over time or within the
sciences. Even if variables such as GNP turn out to be roughly correlated with a country's scientific output over a long historical period,
this correlation might not explain variations in national prominence
at particular periods of time. Qualitative historical analysis has led us
to believe that England was the home of the scientific renaissance of
the seventeenth century; Germany the scientific leader in the second
half of the nineteenth century; and the United States the dominant
country in science since the end of World War II. Few sociological
studies have been conducted on the causes of the rate of scientific
advance.

To what extent is the rate of scientific advance in a given society at
a given time a function of the number of people who become scientists, and what social factors influence the flow of talent into an occupation like science? The theory I am exploring states that the rate of
scientific advance in a given society at a given time will depend upon
the number of talented people who choose science as a career. Scien-

tific advance is uniformly defined as the number of significant scientific discoveries made. Measuring what is an "important" discovery and counting them is a tricky methodological problem which will be discussed below.

To find that the rate of scientific advance is indeed dependent upon the number of people entering science would be only a first step. We would then have to specify the social and cultural conditions that determine the number of talented people who enter science in a particular society at a particular time. Price's analysis of the role of GNP and of economic development as measured by power consumed is important, because it suggests certain types of variables which might influence how many talented people go into science. Although Price says that there is a stronger correlation between the number of scientists and GNP than there is between the number of scientists and the amount spent on higher education, he never provides us with any data on the latter variable. GNP is undoubtedly related to the proportion of a population who receive a higher education, a prerequisite for becoming a scientist. GNP might also influence the level of societal resources spent on science. Although further research is clearly needed on this topic, it is probable that economic development places only broad limits on what proportion of talented individuals in a population will become scientists.

Cultural and Structural Theories

Sociologists have used two general explanations of scientific advance: cultural and structural. The cultural explanation looks at choice of science as a career as a function of the value system of the society. Some societies, it is argued, as a result of their culture, place a higher value on scientific and technological activities than do others. The best example of this type of research is Merton's *Science, Technology, and Society in Seventeenth-Century England* ([1938] 1970).

Most sociologists of science, including Merton, have at least implicitly assumed that the amount of potential scientific talent does not vary from society to society or within a society over time. They have generally assumed a scarcity of "developed" talent in all societies, and they have focused on the distribution of talent within a society rather than on the total amount of talent. Presumably England was not the preeminent country in science in the seventeenth century because the English in that period were more intelligent than the French or the

Germans. Rather, Merton argued the scientific success of a society depends upon its ability to channel more of its available talent into science:

> The Carlylean "heroic" explanation purports to find the origin of periods of intellectual efflorescence in the simultaneous appearance of geniuses. But, as has frequently been observed, in considering the periods in which an unusual number of intellectual giants appear, the phenomenon to be explained "is perhaps not the multiplication of superior natural endowments but the *concentration* of superior endowments upon the several occupations concerned." . . . The more plausible explanation is to be found in the combination of sociological circumstances, of moral, religious, aesthetic, economic and political conditions, which tended to focus the attention of the geniuses of the age upon specific spheres of endeavor. A special talent can rarely find expression when the world will have none of it. ([1938] 1970, p. 5)

Merton's work is based upon the assumption that the various social institutions compete for the limited supply of talent available at any given time. In order to study the flow of talent over time in England empirically Merton conducted a quantitative analysis of the entries in the *Dictionary of National Biography (DNB)*. Merton notes that the number of scientists listed in the *DNB* may not be an accurate indicator of the importance of science as compared with other areas of human endeavor in the seventeenth century; the values of the compilers of the *DNB* could well have influenced the relative number of people included in the various fields. But he argues that because the *DNB* was put together at one point in time, the values of the compilers remain constant for the century he is interested in. Therefore, if an increase in the number of scientists over the century is found, it is probably a result of increasing interest in this field. For each segment of the seventeenth century Merton counted the number of people listed in the *DNB* (defined as a sample of the talented people) entering the various occupations. He found a significant increase in the proportion of talented people entering science in the third quarter of the seventeenth century.[5] He traces this increase in large part to the emergence of ascetic Protestant religious values, which he sees as placing a higher value on science than did Catholicism:

> The social values inherent in the Puritan ethos were such as to lead to an approbation of science because of a basically utilitarian orientation, couched in religious terms and furthered by religious authority. Scientific investigation, viewed from the rationalized Puritan system of ethics,

seemed to possess those qualities characteristic of activities which are effective means for the attainment of the accepted goals. ([1938] 1970, p. 79)

And later:

Empiricism and rationalism were canonized, beatified, so to speak. It may very well be that the Puritan ethos did not directly influence the method of science and that this was simply a parallel development in the internal history of science, but it becomes evident that, through the psychological sanction of certain modes of thought and conduct, this complex of attitudes made an empirically founded science commendable rather than, as in the mediaeval period, reprehensible or at best acceptable on sufferance. In short, Puritanism altered social orientations. It led to the setting up of a new vocational hierarchy, based on criteria which inevitably bestowed prestige upon the natural philosopher. (p. 94)

"Culture and values" is thus one of the broad variables determining choice of a scientific career.

Merton's thesis has been widely debated.[6] Unfortunately, from a sociological point of view, there has been too much concentration on whether Merton's *specific* conclusions about seventeenth-century science are correct and little attention paid to the more general theory underlying his explanation. The underlying theory, never further developed by Merton, states that the amount of talent entering science varies with the cultural value a society places on science. The project of testing and elaborating the theory through broad cross-cultural and historical analyses in which both the value placed on science and the rate of advance are measured remains on the research agenda.

The structural explanation of scientific advance sees the number of talented people entering science as a result of opportunity structure.[7] The more jobs that are available in science, the more people will enter science. The sociologist most associated with this approach is Joseph Ben-David. Ben-David and Zloczower (1962) conducted a study of the emergence of physiology as a science in the second half of the nineteenth century. Interested in what countries most contributed to this emerging field, they counted the number of discoveries listed in history of physiology texts. They found that in the last half of the nineteenth century Germany was the leader in physiological research as in many other scientific areas, and they trace the success of German science to the structure of the German university system. France and England had one or two major universities that dominated scientific life, but in Germany there were roughly twenty universities which competed with each other for scientific prominence. This competition created an expanding opportunity structure for scientists. When one university established a chair in a particular specialty, such as physiology, the other universities would follow suit in order to remain competitive. In a relatively short period of time there was thus a significant increase in the number of academic job openings. These openings attracted talent into the new fields. Ben-David's work is based upon the implicit assumption that scientific advance will be influenced by the number of jobs available for research scientists, and that the universities, as a major source of such jobs, play a significant role in the creation of scientific opportunity structures.

Although Ben-David counts the numbers of discoveries made in different countries, his analysis is essentially qualitative. He does not have any measures of the actual number of scientific jobs available in different countries over time. In the German university system on which he bases most of his analysis, only about twenty new chairs of physiology were created over a period of twenty-five years. This number raises the question of how talented youth knew that opportunities were opening up and how the relatively small number of opportunities opening up over a twenty-five-year time period could have been responsible for the outcome Ben-David and Zloczower observed. Ben-David never raises the question of how the total number of people entering science is related to the level of progress; he implicitly assumes that it is a linear function.

Measuring the Rate of Advance

Both the work of Merton and Ben-David make the implicit assumption that the number of creative scientists capable of making a significant contribution will be a direct function of the total number. In order to explore this hypothesis we must be able to measure the rate of advance and to define a "significant" contribution. These problems are among the most difficult in the sociology of science. When we study the growth of scientific knowledge we face a potential problem of distinguishing between the quantity of knowledge and the "quality" of knowledge. If the growth of knowledge or scientific advance is conceptualized in quantitative terms, then the measurement problem becomes manageable. After defining the universe of scientific journals, we can count the number of articles published or the number

of articles receiving a minimum number of citations. But most scholars who have studied this topic define knowledge not simply as the
accumulation of papers but as the accumulation of "important" discoveries. Certainly *knowledge,* as opposed to science, is growing faster
in a period when relatively few papers are published but among them
are important discoveries than in a period when many papers are
published but few important discoveries are being made. Thus a distinction might be made between the growth of frontier knowledge
and the growth of core knowledge.

Consider first the simpler problem of determining what the most
important discoveries are in a given research area at a given point in
time. Most scholars in the social studies of science today would agree
that there is no objective or absolute way to determine this. One
must therefore depend upon the subjective assessments of those most
qualified to judge: the other scientists working at the research frontier. The most important discoveries in a given research area at a
given point in time could thus be defined as those receiving the most
recognition, whether measured by citation, receipt of awards, or
other indicators. Such a procedure today is nonproblematic. The
problem arises when we want to count important discoveries in such
a way that discoveries made in different times and places can be
compared.

Sorokin and Merton (1935) analyzed this problem using the history
of Arabian science as a research site. They experimented with different methods of measuring the amount of progress made in different
time periods. Initially they counted the number of discoveries made
in Arabian science from 700 to 1300. They then weighted each discovery by the amount of space devoted to it in Sarton's ([1927] 1962)
Introduction to the History of Science in order to make an approximate
measure of the significance of the discovery. Sorokin and Merton
found that these different techniques of measuring the dependent
variable yielded essentially the same results. At least for Arabian me-
dieval science, the total number of discoveries deemed important
enough by Sarton to merit inclusion was correlated with the importance of the discoveries. Later, in *Science, Technology, and Society in
Seventeenth-Century England* ([1938] 1970) Merton drew the following
conclusion:

> Broadly speaking, there seems to be an appreciable correlation between
> the *number* of workers and discoveries in any given field and the *impor-*

tance of both the scientists and discoveries . . . The correlation is probably due to the fact that outstanding scientists frequently attract a cluster of less talented followers, so that periods with an unusually large number of brilliant scientists are also periods of great interest in science. Moreover, conspicuous success in any given field . . . is apt to attract the attention of a number of mediocre as well as able investigators. (p. 40, note 4)

If it is true that the number of important discoveries is highly correlated with the total number, we could count the total number of discoveries and assume that this indexed the number of important ones.[8] Other researchers, however, have not found the same results that Sorokin and Merton did.

Price's Theory of Scientific Growth

The assumption that important discoveries will be a direct function of the total number of working scientists was challenged by Derek J. de Solla Price (1963) in *Little Science, Big Science*. Price begins with the observation that scientific productivity is very highly skewed, with a small proportion of scientists being responsible for the great bulk of scientific papers. He points out that about 10 percent of scientists publish about 90 percent of the papers. He goes on to suggest that although the total number of scientists has been growing exponentially for a long period of time, the number of outstanding or creative scientists has been growing at a considerably slower rate. He concludes that if we want to multiply the number of "good" scientists by a factor of 5 we must multiply the entire group of scientists by a factor of 25 (p. 53). This theme recurs throughout the book.[9] He argues that whereas the doubling time for all scientific work is about ten years, the doubling time for "very high quality" work is about twenty years.

What evidence did Price have to support this conclusion? The only evidence cited is the rate of increase in "starred" men in *American Men of Science* (see Price, 1963, pp. 36–37) and the rate of increase in Nobel Prize winners (pp. 39, 56n). But, as Price is aware, these figures are an artifact of the designation procedures. Because there is only one Nobel Prize awarded in physics each year (ignoring shared prizes), the rate of increase in Nobel Prize winners must, of course, be linear, no matter what the rate of increase in the overall size of science.[10]

Price discusses two ways of determining who has made a major contribution to science:

> One may set a limit and say that half the work is done by those with more than 10 papers to their credit, or that the number of high producers seems to be the same order of magnitude as the square root of the total number of authors. The first way, setting some quota of 10 or so papers, which may be termed "Dean's method," is familiar enough; the second way, suggesting that the number of men goes up as the square of the number of good ones, seems consistent with the previous findings that the number of scientists doubles every 10 years, but the number of noteworthy scientists only every 20 years. (1963, p. 46)

The "previous findings" Price refers to are the data mentioned above. Price may have found the second way more intuitively correct, but because he never suggests that publishing ten papers today is less of a contribution than publishing ten papers in the past, his choice is questionable. In fact Price frequently uses number of papers published (and later, number of citations received) as a historically independent indicator of contribution to science. If "Dean's method" is used to determine the number of "high-quality" scientists, then there is no reason to expect the proportion of such scientists to decline as the total number grows.

In order to understand Price's assumption about the rate of increase of good scientists, we must examine his largely implicit underlying theory of what influences the number of good scientists. Price follows Galton here and assumes that in any population there are a limited number of talented people who are capable of becoming scientists. Price defines "talented" as having an IQ of at least 130. Of these he assumes that a disproportionate number will choose to enter science. Therefore an increase in the size of science will necessarily result in a smaller *proportion* of talented people entering the field, because there will be fewer talented people left who have not already chosen science as a career.

Price makes this theory clearest when he discusses why we can expect the fastest growth in science in countries that are currently underdeveloped. In the developed countries a high proportion of those capable of doing science already select that career. In the developing countries there are greater potential reserves of talent. Price's analysis sees scientific talent as fixed and assumes that the most talented will be drawn into science. His assumption that there will be a declining proportion of good scientists as the total number increases is based upon his belief that the average IQ or scientific ability of those who enter science will have to decline as the numbers entering increase.

Rescher's Theory of Scientific Progress

The philosopher of science Nicholas Rescher has elaborated some of Price's ideas in a series of essays on scientific progress. Rescher (1978) agrees with Price that, as the total number of people entering science increases, the proportion of those making truly important discoveries will decline: "For the rapidly—indeed exponentially—increasing pace of effort-investment tends to mask the fact that the volume of high-quality returns per *unit* investment is apparently declining" (p. 87). Unfortunately, Rescher has no more empirical evidence to back up this claim than did Price. The evidence he refers to is the increase in the number of scientists receiving prizes, medals, and membership in prestigious scientific organizations and the listing of discoveries in "synoptic handbooks, monographs and treatises that endeavor to give a rounded picture of the 'state of the discipline' " (p. 89). As I pointed out above, the number of people receiving Nobel Prizes must increase linearly because of how the prize is distributed. Rescher does not consider the ways in which using handbooks may introduce the positivistic bias of the compilers. Compilers of such works probably rely heavily on core science. Because at any given point a large portion of all science is recent science and because much important recent work is probably still on the frontier, it is quite likely that the relative significance of recent work could be substantially underestimated by this method of measurement.

Although Rescher and Price both believe that science encounters the problem of diminishing returns, their reasons differ. Rescher sees limits placed not on the availability of talented people but on the availability of economic resources necessary to do science. The number of scientists has been growing at an exponential rate, with a doubling time of about twelve years, but the cost of doing science has been increasing at an even faster rate: "Our deceleration-thesis rests neither on the claim that the well of undiscovered significant findings is running dry, nor on the claim that the limits of man's capacities for inquiry are being exhausted. The slowing it anticipates is not *theoretically inevitable* for any such absolutistic reason, but rather is *practically unavoidable* for the merely *economic* reason that it is getting more and more expensive to run the increasingly complex machinery of scientific innovation" (1978, p. 121). Rescher suggests but does not take advantage of a research site which might enable the testing of his hypothesis. He points out that some fields, such as mathematics, do not require new data and therefore should not encounter the deceleration resulting from resource requirements. The rate of progress in mathematics, but not in physics, should therefore be a result of the number of people entering the field.[11] But without such data Rescher's analysis is completely dependent upon his assumption that the number of important discoveries is not a linear function of the total number made.

Rescher makes the interesting argument that although it becomes more and more difficult to make discoveries that are truly important, the distribution of recognition in the scientific community is unlikely to change. Those scientists whose work is deemed the best at any given time will be anointed as stars, even if the stars of today are not doing work comparable to the stars of yesterday.

> To be sure, the theory of deceleration contains the prospect of longer and longer periods of elapsed time between the realization of really first-rate discoveries—a situation of continually increasing waiting times for equal progress. But all this is by *absolute* standards, while men in fact tend to make their judgments on a relative and comparative basis. We incline to esteem as first-rate the best findings realizable under the working conditions of the day, esteeming scientific greatness by the relativistic standards of the researcher whose work figures massively in the citation-indices and whose labors are held up as latter-day exemplars to his colleagues. The natural human tendency is to construe "the best" as "the best that can be expected in the circumstances" . . . The age of heroes will not have come to an end: the heroes of the day will still be those who win its races. (1978, p. 260)

Does it make sense to try to compare the significance of scientific discoveries made at different times? Rescher assumes that the laws of nature are external to the discoverers and that at any level of understanding the number of these laws is finite. At any level of analysis, therefore, as important laws are discovered it will be more and more difficult to discover new equally important laws—simply because fewer undiscovered laws will exist at that level. To make equally important discoveries it will be necessary to go to a new level of analysis, requiring more expensive data collection and analysis instruments. If we accept the Kuhnian view of science, which changes our notion of scientific progress, it becomes difficult to make sense out of the effort to compare the importance of discoveries made at different points in time. Are discoveries and whole programs of research which turn out later to be "wrong" to be counted as progress? Rescher agrees that scientists evaluate work relative to other work done *at the same time,* but who is to make cross-temporal evaluations of the importance of work? Rescher would leave this task to the compilers of history books. But to do so seems unsatisfactory for all the reasons upon which Kuhn and others base their criticism of the traditional history of science. It is also unlikely that historians of science who accept the new philosophy and its corresponding historiographical methodology would be willing to make judgments about the absolute importance of discoveries made in different periods. Although I do not reject the idea that it might be *conceptually* possible to define criteria of "importance" and then make distinctions among discoveries, I believe that this approach is empirically very difficult or impossible and therefore represents a blind alley for the sociology of science.[12] Rather than attempting to compare the significance of important discoveries made in different eras, the sociology of science should concentrate on finding those conditions which influence the production of important discoveries as they are defined by contemporaries. Even if "relativistic" criteria of defining important discoveries are used, it remains an interesting question whether the number of important discoveries so defined is, as Merton and Ben-David implied, a linear function of the total number of scientists or, as Price argued, a squared function of the total number of scientists.

The Ortega Hypothesis

Influenced by the work of Price, Jonathan R. Cole and I in the early 1970s investigated the hypothesis that it might be possible to reduce the size of science *without* a corresponding decline in the quality of science. It seemed clear on the basis of the distribution of publications and of citations to these publications that only a small portion of all scientists actually contribute to scientific advance through their published research. It was not clear, however, to what extent the small number of scientists who make the significant discoveries depend upon the large number of scientists who do less significant work. The research we conducted, published in a paper entitled "The Ortega Hypothesis" (J. R. Cole and S. Cole, 1972), aimed at discovering the extent to which scientific advance depends upon the work of all "social classes" or is primarily dependent on the work of an "elite." This hypothesis was tested in several different ways. In general, we found that the great bulk of work cited in important discoveries is the work of other eminent scientists. We concluded that the giants of science stand not on the shoulders of a mountain of dwarfs but on the shoulders of a few other giants.[13]

This conclusion, which is virtually predetermined by the marginal distribution of productivity, has been accepted by most sociologists of science.[14] Considerably more controversial has been the policy conclusion we drew from the research. We concluded that it might be possible to reduce significantly the size of science by reducing the number of graduate students admitted and eliminating doctoral programs at weaker universities without slowing down the rate of scientific advance. This policy suggestion was based upon several assumptions similar to those made by Price. First, we assumed that the brightest, most highly motivated undergraduates would continue to want disproportionately to become scientists. Second, we assumed that it would be possible for graduate schools to distinguish between those applicants with the most potential and those with less potential. Thus, although the total number of graduate students might decline, the number of the most talented students would either not decline at all or decline at a significantly slower rate.

In 1972, when "The Ortega Hypothesis" was published, there was no consensus among sociologists of science on the question of whether the number of important scientific discoveries is a linear function of the total number of scientists. The work of Merton and Ben-David adopted this as an implicit assumption from which it would follow that a reduction in the number of scientists would bring about a reduction in scientific creativity. Price and Cole and Cole made the opposite assumption.

Empirical Evidence on Physicists

In the late 1970s Garry Meyer and I conducted a study that sought to find out which of these two assumptions was correct. In 1963, when Price published *Little Science, Big Science,* the size of science had been rapidly expanding. Price knew that this rate of increase could not go on forever and accurately predicted a leveling off of the logistic curve. It is not clear, however, whether Price could have known how quickly that decline would occur. As one indicator of the rate at which science was growing in the United States, Price used the number of entries in successive editions of *American Men of Science*. In the 1965 edition there were 130,000 entries, or 642 per million of U.S. population. In the 1982 edition of *American Men and Women of Science* there were still only 130,000 entries, or 560 per million of U.S. population. It is evident that since the time of Price's book, at least in terms of this indicator, there has been little growth and certainly no exponential growth in the size of American science. Owing to demographic factors (essentially the end of the baby boom and a decline in the rate of increase of students enrolling in college), a sharp reduction in the demand for academic scientists began in the late 1960s. To what extent has the slowdown in the growth of science (and indeed an absolute decline in the number of new Ph.D.s) had an influence on the number of creative scientists? The dependent variable for this analysis is the absolute number of scientists who make significant discoveries. The independent variable is the total number of scientists. Presumably because fewer academic jobs were available, fewer people applied to graduate school, earned doctoral degrees, and became science professors.

The purpose of the Meyer and Cole study was, first, to measure the extent to which there has been a decline in demand for American academic scientists in the field of physics and, second, to assess the effect that this decline has had on the number of new physicists who will be capable of making significant contributions to knowledge.[15] Although there had been considerable research devoted to projecting future labor markets for Ph.D.s, there had been only speculation on how changes in demand for scientists would affect the quality of science produced (Klitgaard, 1979).

It is now clear that there has in fact been an overall decline in the demand for scientists since the 1960s (Breneman, 1975; A. Cartter, 1976; Radner and Miller, 1975). Most of these studies, however, treat scientists in a particular field as a group and do not differentiate demands for different types of scientists. Lumping together scientists who work in Ph.D.-granting institutions, other educational institutions, government laboratories, and private industry may obscure the change in demand which has occurred for the type of scientist who is primarily responsible for the growth of basic scientific knowledge. In the United States in most fields basic knowledge is primarily produced by individuals who have academic appointments at major universities. In 1976, 68 percent of all published research in physics was produced by scientists employed in colleges and universities, and most of these worked in research universities. The case study measured the strength of demand for academic physicists employed at American Ph.D.-granting universities between 1963 and 1976.

Grodzins (1976) has shown that in the early 1960s the total number of Ph.D.-granting departments in physics substantially increased. The size of the faculties of these departments grew rapidly until the end of the decade, when the trend reversed. To study the demand for young physicists, which was not adequately indicated by the total number of faculty employed, we counted the new assistant professors hired by Ph.D.-granting departments in every year from 1963 to 1976.[16] The total number of new entrants to the "system," defined as the group of all Ph.D.-granting departments in physics, indicates that there was an increasing demand for young academic physicists through the 1960s which peaked in 1966 (see Column 1 of Table 9.1). Demand for academic physicists then declined so that by the year 1975, the size of the entering cohort of new assistant professors was less than half the size of the peak year. For physics, at least, this period witnessed first a sharp growth in size (similar to the exponen-

Table 9.1 Data on citations to new cohorts of physicists, 1963–1976

Cohort	(1) Size	(2) Number receiving 1+ citations	(3) Percent receiving 1+ citations	(4) Number receiving 5+ citations	(5) Percent receiving 5+ citations
1963	222	57	26	15	6
1964	283	92	32	36	12
1965	337	129	38	50	14
1966[a]	406	114	28	27	6
1967	369	127	34	38	10
1968	374	116	31	32	8
1969	315	113	36	31	9
1970[a]	284	77	27	21	7
1971	230	62	27	18	7
1972	138	43	31	5	3
1973	175	49	28	17	9
1974	176	63	36	28	15
1975	155	51	33	11	7
1976[a]	149	46	31	12	8

Note: For the 1963–1965 cohorts the 1967 *SCI* was used; for the 1966–1970 cohorts the 1971 *SCI* was used; for the 1971–1976 cohorts the 1977 *SCI* was used.

a. Based on only 2 years' publication.

tial growth Price analyzed) and then a sharp decline, which I take as indicating a drop in demand for young academic physicists.

We also compared the scientific output of those physicists who entered the system in 1963 with those who entered in all subsequent years through 1976. (The unit of analysis for this study was the entering cohort of new faculty members in each year.) The output measure used was the number of citations received by the members of the cohort in the first few years after entrance. We compared the scientific output of each cohort produced within one, two, three, and four years of entering the system. We wanted to determine both the absolute number and the proportion of each entering cohort who had made a significant contribution during the first several years after entrance to the field.

Although I concluded in Chapter 6 that the "quality" of scientific work cannot be objectively assessed, extensive past research indicates that citations are a valid indicator of the subjective assessment of quality by the scientific community. The number of citations is highly correlated with all other measures of quality that sociologists of science employ. As long as we keep in mind that research of high quality is being defined as research that other scientists find useful in their current work, citations provide a satisfactory indicator.[17]

Because the cohorts studied were relatively young, we were unable to assess their lifetime contributions. Early contributions to science, however, are an important indicator of future contributions. Past research has shown strong correlations between the early and later productivity and creativity of scientists. Although some scientists who begin their careers by being creative and productive experience a gradual reduction in their scientific output, it is infrequent to find examples in which scientists begin their career by being nonproductive and later become highly productive (S. Cole, 1979). In a study of a cohort of about 500 mathematicians who received their Ph.D.s in the United States between 1947 and 1950, Nancy Stern and I (S. Cole, 1979; Stern, 1978) found that only a small handful were productive later in their careers and unproductive early in their careers.

Meyer and I were able to use the data on physicists hired between 1963 and 1965 to examine the relationship between the quality of work produced in the first several years after being hired and the quality of later work. A total of 842 physicists were hired as new assistant professors in the years 1963, 1964, and 1965. We had data on the number of citations received by their work published in four different periods: (1) the year in which they were hired plus the two following years, (2) the third through the fifth year of their employment, (3) the sixth through the eighth year, and (4) the ninth through the eleventh year. These data indicate that there were very few physicists who started their careers without producing cited work and who were later able to produce cited work (see Table 9.2).[18] Of new assistant professors hired between 1963 and 1965, 376, or 45 percent, did not receive a single citation throughout the eleven-year period. Only 212, or 25 percent, received five or more citations to work published in one or more of the four periods. In general, those physicists who begin their careers by publishing work which is cited are far more likely than those whose early work is not cited to continue to produce high-impact work. Of the 692 physicists who received fewer than five citations to their work in the first three years after entrance, only 1 percent received five or more citations to work published six to eight years after entrance and 2 percent received five or more citations to work published nine to eleven years after entrance.

Even if we use the less exacting criterion of having received only one citation to their work published in the first three years after entrance, we find that among those with zero citations to early work only 10 percent received one or more citations to work published six to eight years after entrance and only 7 percent received one or more citations to work published nine to eleven years after entrance. We

Table 9.2 Citations to early and later work published by new physicists

Number of citations to work published in first 3 years	Percentage receiving 5 or more citations to work published		
	6–8 years after entrance	9–11 years after entrance	N
5 or more	23	19	150
Less than 5	1	2	692
	Percentage receiving 1 or more citations to work published		
	6–8 years after entrance	9–11 years after entrance	
1 or more	38	32	333
0	10	7	509

concluded that although there are some scientific "late bloomers," in general, early productivity is a good rough predictor of later productivity. Indeed, because there are many more scientists whose productivity declines than there are those whose productivity increases, the measures we used in this study may overestimate the total number of scientists who will be significant contributors to new knowledge in the future.[19] Thus by determining the number within each entering cohort who do creative work in the first several years after entrance, we can project a "maximum" estimate of the number of scientists who will be available in that cohort to produce new knowledge. Remember also that we were not interested in predicting which members of a cohort would be productive, but only in predicting the absolute and relative size of the productive group. Errors introduced by our measurement procedure should be random, and should not affect the substantive conclusion.

My primary concern is the relationship between the overall size of each new cohort and the number of creative scientists in the cohort. If Price is correct we should expect to see the proportion of creative scientists decline as the size of entering cohorts grew in the 1960s and then increase as the size of entering cohorts decreased. The *number* of creative scientists should remain roughly constant over the entire time period. If the implicit assumption of Merton and Ben-David is correct, we should expect that the proportion of creative scientists would be a direct function of the total number. If this is true the absolute number of creative scientists should decline as the total number declines.

Many difficult methodological problems had to be dealt with in measuring the scientific output of the physicists in the study. Throughout the period of time under consideration, the size of the *Science Citation Index* files was increasing. The 1961 edition of the *SCI* was based on somewhat more than 600 journals; the 1991 edition of the *SCI* is based on information extracted from more than 3,000 journals. Because of this increase in the size of the *SCI* files, there has been a corresponding increase in the number of citations received by scientists. To take this into account we looked at citations made only in the two leading physics journals, *Physical Review* and *Physical Review Letters*. Citation data show that these two journals are the dominant journals in the field of physics and represent a significant proportion of total citations made in the field of physics (Narin, 1976). They have also been included in the *SCI* files from the beginning.

Even considering only these two journals between the years 1967 and 1977, however, there was a significant increase in size and the total number of citations made.[20] In parts of our analysis it was therefore necessary to express the data in 1967 standardized units. A more serious problem concerns the relationship between citation patterns and the age of literature. On *average,* physics papers receive their highest level of citations one year after they are published. For each year that the paper ages the average number of citations received goes down sharply. It is thus difficult to compare the citation level of work of different ages. This problem must be taken into account in evaluating the meaning of the results reported below.[21]

Let us begin by examining the proportion of each cohort of new assistant professors whose work, published at any time during the first three years after initial employment as an assistant professor, received at least one citation. It can be argued that such a criterion is inadequate if we are interested in the number of scientists who have made or will make truly significant contributions to physics. This is certainly true, but it should be kept in mind that given the short time span over which we are evaluating the output of the young physicists it is impossible to come up with the number of each cohort who will make such significant contributions. What we are trying to do is to estimate the *maximum* number likely to engage in research which some other scientists will find useful. The number who will ultimately make significant discoveries will undoubtedly be substantially lower than the number who will have work cited at least once.[22] In the citation analysis we have used the citation index which was closest in time to the period being considered. Thus, for the 1963–1966 cohorts we used the 1967 *SCI;* for the 1967–1970 cohorts we used the 1971 *SCI;* and for the 1971–1976 cohorts we used the 1977 *SCI.*[23] In the 1963 cohort, 57 members published work between 1963 and 1965 that received at least one citation; in the 1965 cohort, 129 members published work between 1965 and 1967 that received one or more citations; and in the 1972 cohort only 43 physicists published work in 1972 to 1974 which received one or more citations (see Column 2 of Table 9.1). The *proportion* of each entering cohort whose published work received one or more citations in the first three years after entrance ranges from 26 to 38 percent and overall is fairly constant at approximately 30 percent (see Column 3 of Table 9.1).

Using the same data we counted the number of physicists in each cohort whose published work was cited five or more times within the first three years of their initial employment as an assistant professor (see Column 4 of Table 9.1). Although receiving five or more citations does not necessarily indicate that the physicist had made a "truly significant" discovery in Rescher's terms, very few physicists have received this many citations to early work. About 9 percent of all the young physicists in this study had received this many citations to their early work (see Column 5 of Table 9.1). Fifty members of the 1965 cohort and only 5 members of the 1972 cohort received five or more citations. Although the proportion of each cohort to receive five or more citations is not quite as stable as that reported in Column 3 of Table 9.1, we can see no evidence either that as the size of the cohorts grew larger the proportion of highly cited physicists grew smaller or that as the size of the cohorts grew smaller the proportion of highly cited scientists grew larger. The variation in the proportion of highly cited scientists seems to follow no pattern. There are some years, such as 1965 and 1974, which have particularly high proportions of highly cited scientists. Research is needed on whether these findings result from the random distribution of talented young scientists across time or can be traced to particular developments in physics.[24]

The physics results show that in a field that has experienced first a sharp rise and later a sharp reduction in the size of its entering cohorts, there has been a corresponding rise and reduction in the *number* of scientists making early contributions to knowledge in their field. The demand for academic physicists working in Ph.D.-granting departments peaked in 1966 and fell off sharply in the early seventies. As the size of new entering cohorts to the system of Ph.D.-granting departments declined, there was a corresponding decline in the number of new young physicists who made significant discoveries in the first three years after entering the system. This suggests that contrary to the assumptions of Price (1963) and J. R. Cole and S. Cole (1972), the number of scientists who will make significant contributions to science *is* in general a linear function of the total number of people entering science.

The data presented in this chapter offer empirical support for the implicit assumption of the work of both Merton and Ben-David that scientific advance is a function of the number of people entering science. The data do not support the assumption of Price (1963) that the number of good scientists grows at a slower rate than does the total number of scientists; nor does it support the assumption of the Cole and Cole (1972) Ortega hypothesis paper that the size of science could be reduced without reducing the rate of scientific advance.

Conclusion

I would speculate that the most important reason for the observed decline in the number of talented young physicists is the significance of opportunity structure in influencing the career choice of talented individuals. People who select science as a career are obviously those who have an intense interest in the subject. At least in our times, however, intense interest does not seem to be enough. Many potential creative scientists are discouraged by the lack of secure employment prospects. As the job market for academic scientists has tightened, the attractiveness of the occupation to talented as well as less talented youth has decreased. Whereas in the 1960s talented young people may have selected the sciences in preference to the other professions or business, that trend seems to have been reversed.

One way to examine this hypothesis would be to look at the career choices of the top 10 percent of graduates of highly selective colleges over time. Are students who have done the best in college less likely to choose science as a career today than they were at other times in recent history? Some preliminary data suggest that this is indeed the case. If we examine the number of National Merit Scholars (the most prestigious scholarship offered to high school students in the United States) selecting various college majors we find a precipitous decline in the proportion selecting physics between 1966 and 1976 and a corresponding increase in the proportion selecting a premedicine program (NSF, 1977, p. 288). In 1966 about 8 percent of National Merit Scholars said they would major in physics, but by 1976 this had fallen by 50 percent to about 4 percent of Merit Scholars. The percentage saying they were premedical students increased from about 5 percent in 1966 to 11 percent in 1976, more than a 100 percent increase.

Studies similar to the one Meyer and I conducted should be repeated on a broad range of scientific fields in as many societies and time periods as possible. Can we identify a general uniformity of scientific advance? Assuming that we can establish that the rate of advance is correlated with the number of people entering science, this correlation is bound to be far from perfect and to raise the next set of important intellectual problems. Under what social and cultural circumstances is scientific manpower converted into scientific advance? Particularly interesting are variables characterizing the social organization of science. The way in which science is structured varies dramatically. The organization of science in Western Europe, for example, differs in many important respects from that in the United States. Can variables be developed characterizing the forms of social organization and can we then examine the correlates of varying forms of social organization and rates of scientific advance? Although a successful investigation of this type is many years in the future, the policy implications of such a study should be clear. We know very little about why talented people enter science, medicine, or other careers. Yet it is possible that the ability of each of these occupations to serve its socially assigned function is in part dependent upon this flow of talent. It is a question worth further pursuit.

The Future of the
Sociology of Science

Is the growth of scientific knowledge influenced by social variables? My answer is a highly qualified yes. This response can be summarized by returning to three important conceptual distinctions made in this book: the distinction between the foci of attention and the rate of advance on the one hand and the actual substance of ideas on the other, the distinction between a local knowledge outcome and a communal knowledge outcome, and the distinction between core knowledge and frontier knowledge.

Most observers of science are in agreement that the foci of attention (the choice of problems to work on) and the rate of advance (the amount of new knowledge developed) are influenced by social variables. Scientists are more likely to work on scientific problems for which the society is willing to provide resources. The way in which social, political, and economic forces interact with the interests of the scientific community in setting the research agenda is an important problem requiring much additional research. But it is already clear that social variables will play at least some role in the selection by scientists of particular research problems and thus in which problems are most likely to be solved. In addition, as the data in Chapter 9 suggest, the rate at which scientific knowledge grows is at least in part determined by attributes of the culture and social structure which influence the flow of talent into science.

Most of this book has concentrated on the extent to which social variables influence the actual cognitive content of scientific ideas as these ideas come to be accepted by the scientific community. The dominant view in the sociology of science today on this question is that of the social constructivists. They believe that the substantive content of scientific theories is socially determined or constructed.

Each new scientific contribution is developed by one or more scientists working in a local context. A local knowledge outcome is the finished product of this effort, usually a published scientific paper. The first problem we must review is the extent to which the substance of local knowledge outcomes is influenced by social variables and processes. There can be no doubt that the construction of local knowledge outcomes is carried out in a social environment and is therefore influenced by social processes. Thus two scientists will engage in negotiation over sharing resources, their decisions may be influenced by career interests, and the operation of intellectual authority may determine whether one view is accepted over another. But does the fact that the doing of science is a social activity mean that the substance of the local knowledge outcome is determined by this activity? Unfortunately, even for this micro-level question, the studies carried out by social constructivists do not provide an adequate answer. Frequently these studies black-box the cognitive content or fail to show a specific relationship between a social activity and the substance of the local knowledge outcome. Constructivist studies do show that the doing of science is not the rational rule-governed activity it has been depicted as and that serendipity and chance play a significant role in the construction of local knowledge outcomes. Studies done by social constructivists do suggest (but have not yet demonstrated) that local knowledge outcomes *may* be influenced by social variables. These studies have *not* proved that the extent to which theories match data from the empirical world has *no* influence on local knowledge outcomes. They show that science is underdetermined but do not show that it is totally undetermined.

If we grant that local knowledge outcomes are influenced, in part, by social variables and processes, We are left with the question of the extent to which communal knowledge outcomes are similarly influenced. I have divided knowledge into two major components: the core and the frontier. The core is the small body of knowledge that the community of science accepts as both true and important. A contribution becomes a communal knowledge outcome when it is accepted into the core. The frontier consists of all knowledge produced by community members. Only a very small portion of new knowledge moves from the frontier to the core.

The substance of new ideas is determined in the local context.[1] The evaluation process does not affect the substance of new ideas but determines which among different substantive contributions are accepted as true and important by the community. The social processes involved in evaluation therefore influence the foci of attention of the scientific community but not the actual substance of ideas admitted to the core. Thus the main social variables analyzed in this book, the characteristics of scientists making the contributions and the operation of intellectual authority, affect not the cognitive content of core knowledge but the foci of attention of the scientific community.

Now let us consider the determinants of evaluation. I agree with the constructivists that evaluation of new scientific contributions at the research frontier is an inherently subjective process. The large number of qualitative case studies conducted in recent years by participants in the social studies of science, quantitative work conducted by other sociologists, and the data presented here all lead to the conclusion that the level of consensus on frontier knowledge is relatively low in all research areas. Many people, because their image of science is based upon core knowledge as opposed to frontier knowledge, are unaware of this low consensus. They believe that physicists, for example, agree on many important theories and "facts." Indeed, they do; but in the day-to-day practice of science the part of physics that they agree upon is not being evaluated. During normal science scientists do not evaluate the core or paradigm of their discipline. When we look at what they do have to evaluate, new contributions, we find disagreement. Whatever enables these scientists to agree upon the core does not work in bringing about consensus on the frontier.

The meaning of this essential conclusion can easily be distorted. When I say that there is no "objective" way to evaluate new science, I do not mean, of course, that any idea, theory, or set of evidence is as good as any other. If this were so, nobody, including social constructivists, would bother doing any systematic scientific work. Evidence from the empirical world and logical procedures do play a very important role in science. They essentially serve to rule out most *conceivable* solutions, but they frequently fail to distinguish among the *proposed* solutions. For most proposed solutions to current scientific problems it is usually difficult to reach consensus on their importance and validity. Those who do not agree with this position must explain why there is so much disagreement on frontier science.

Because the evaluation of frontier knowledge is inherently subjective, however, does not mean that the cognitive characteristics of a new contribution (the evidence, theories, models, and logic in a new paper) have no influence on the formation of these subjective opinions. It is my view that cognitive characteristics have a significant influence on the evaluation of new contributions, but that the extent of their influence varies widely from one contribution to another. Thus, for some contributions the cognitive characteristics can determine their fate, for others they will play a role in determining their fate, and for still others they may have little or no influence in determining their fate. I would argue, for example, that the fate of the Watson and Crick 1953 *Nature* paper, laying out the model of DNA, was heavily influenced by its content. In this case the characteristics of the authors had little influence, but the process of intellectual authority, although it did not determine the content of the DNA model, served to solidify the importance of the work in the community.

Why does some content compel consensus and admission to the core and other content lead to either disagreement or a passive agreement (through lack of attention) that the paper is not important? I use the concept of utility to answer this question. Those contributions which the members of the scientific community believe to be useful in both raising and solving significant puzzles have a high utility and are more likely to be admitted to the core. The extent to which the content matches evidence from the empirical world is at least one important variable influencing utility. Thus if we consider a local knowledge outcome which has been influenced to some extent by social variables and processes, we have to ask what determines whether this local outcome becomes a communal outcome. My position is that only local outcomes which have high utility will be accepted by the community. The local contextual factors which influence the local knowledge outcome may not influence the communal evaluation process. Ultimately the effect of social variables and processes on the content of knowledge in the core is likely to be quite small.

Essentially my position is a type of realism which sees the actual content of the scientific work accepted as true by a community as being most strongly determined by evidence from the empirical world and by theories or models which attempt to make sense of this empirical evidence. This position is taken not as a result of any philosophical antipathy to relativism but because the analyses of the relativists have been unable to answer the question of why one truth claim meets quick acceptance and most are ignored or rejected.

The problem with the position of the constructivists is that it frequently becomes one of "all or nothing." Either science is determined by empirical evidence and the reception of science is a result of the application of rules and logical criteria, as the positivists would have it, or science is not influenced at all by evidence, rules, and logical criteria, as the constructivists would have it. The constructivists have trouble accepting a more workable position in the middle because they believe that once we acknowledge the nonsocial influences on science there will be no room left for sociology.

Can the Constructivist Position Be Empirically Supported?

Constructivists do not believe that empirical evidence plays a signifi-cant role in evaluation, and therefore the question of whether their position can be empirically supported would be taken by them only as evidence of naiveté. I recognize that in many cases empirical evi-dence does not resolve disputes, and I do not believe that the differ-ences between my position and that of the constructivists will in fact be resolved by empirical evidence. I believe, however, that there have been occasions when the content of a contribution has in fact com-pelled acceptance. Thus it is incumbent upon me to specify what type of evidence would cause me to abandon my position and accept that of the constructivists.

Consider my concept of "utility." It could be claimed that this con-cept is a tautology with no explanatory value. Science which has been successful is defined *after the fact* as having high utility. Just as I chal-lenge the constructivists to define "interests" prior to communal deci-sions, they may raise the same challenge for "utility." Therefore, I shall suggest a way to measure utility prior to communal evaluation of new scientific contributions. Let us consider doing an experiment to discover how cognitive and social variables interact to influence the reception of new scientific work. Suppose that a journal editor agreed to send us manuscripts accepted for publication but not yet actually published. We would then "blind" these manuscripts by removing the name of the author, the institution where the work was conducted, any other identifying information, and even the references in the paper. We would then send this blinded version of the paper out to be read and evaluated by a random sample of authorities (eminent researchers) in the relevant scientific community. After we received the evaluations the paper would be published and we could trace its actual reception by the scientific community. Most of the papers would have little or no impact on the development of communal knowledge; a small number would have a significant impact; and an even smaller number would enter the core and come to be accepted as fact or true. The research question would be to what extent can the evaluations of the readers of the blinded papers predict their reception? I would argue that these evaluations would be a good indicator of the independent effect of cognitive characteristics *and* a way of operationalizing "utility."

What would be the likely results of such an experiment? Some educated predictions can be made on the basis of the peer review study I discussed in Chapter 4 and other work on consensus discussed in Chapters 4 and 5. If these papers were to be evaluated in a way similar to the evaluation of NSF research proposals, we should find both that there is a high level of disagreement among the evaluators and that the evaluations are moderately correlated with later reception. But the level of disagreement about particular papers should vary significantly. For some papers virtually all the reviewers would agree that they were of little scientific value. But on a small group of papers there would be substantial agreement on their importance, just as in the peer review experiment there were some proposals on which there was not significant disagreement. If the decontextualized evaluations of the authorities had little or no power in predicting the future reception of the papers, this finding would be evidence against an important part of my argument and evidence in support of constructivists, such as Latour, who argue that content has no influence on future reception.

Now consider an addition to the experiment. Suppose we also asked the evaluators to answer another question: "Is this work of Nobel Prize caliber?" For the overwhelming majority of articles most evaluators would answer in the negative. For one article in a hundred thousand or a million, we would find significant consensus that it was of Nobel Prize caliber. Although the first part of the study could actually be conducted, this second part can only be carried out as a thought experiment. If the reader agrees that the Watson and Crick article, the Bardeen, Cooper, and Schrieffer article, or the articles by Guillemin and Schally would have received a high consensus on the Nobel question in this experiment, the reader agrees with the position I have taken. If the reader believes there may not have been high consensus in a prepublication evaluation of these articles, the reader agrees with the position taken by the constructivists.

My suggestion that a dispute between myself and the constructivists could potentially be resolved by empirical evidence is reflective of the difference between my orientation and theirs. They would argue that the data would be subject to various possible interpretations and thus would not resolve the issue. I would agree that disputes are frequently not resolved by resort to empirical evidence, but I see this as in part a result of the social organization of scientific research as an activity conducted by only loosely connected individuals or groups of individuals. If scientists who disagreed worked together to resolve the dispute, I am inclined to believe that more disputes could be empirically resolved. In order to work together, of course, scientists must accept the same paradigm. But most disputes among scientists occur during normal science among those who agree upon the core. I wonder whether constructivists would be willing to participate in a joint research effort in which it would be agreed prior to data collection what the significance of various outcomes would be. In other words is there *any* empirical design which constructivists would accept as adequate to compare their interpretation with mine? If the answer is no, then we must wonder why it is that constructivist sociologists bother to do any empirical research and even more why we should pay attention to it. This argument is one of the ways in which relativist approaches to science have been attacked (Laudan, 1977).[2] If we were to carry out a study similar to the one I have proposed and we found that decontextualized evaluations were not correlated with later communal evaluations, this finding would provide negative evidence for my position but would not provide the positive type of evidence required to accept the constructivist thesis that the content of science is determined by social variables and processes.

To devise a study which could demonstrate a causal connection between the cognitive content of science and one or more social variables would be an extraordinarily difficult task. In order to study the causes of a dependent variable there must be variance in that variable. Thus if we wanted to study how social factors influenced the solution to a particular scientific problem, we would have to have different solutions to the problem. The solution rather than the problem should be the unit of analysis. For example, we would have to have different models for the structure of DNA or different sequences for the structure of TRF. But when a problem is solved and its solution becomes a "fact" there is generally only one solution; there is only one model for DNA and only one chemical sequence for TRF. It would therefore be difficult to demonstrate the constructivists' theory using problems which have been solved. We can analyze how social processes were involved in the construction of the fact, but we cannot show that the substance of the fact (the dependent variable) would have been different if the social circumstances (the independent variable) had been different.

Two lines of study, both of which have potentially serious pitfalls, might be pursued. The first possibility would be to study science at the research frontier (as Knorr-Cetina, for example, does), where different solutions to a scientific problem are being proposed. Then it might be possible to show how specific choices or selections are influenced by social processes. Consider, for example, the sixteen drafts of a scientific paper analyzed by Knorr-Cetina (1981). Suppose that we are considering how the negotiation among the different members of the laboratory influenced the final form of the paper. It might be possible to show that the lab director was able to impose his cognitive interests on the juniors who had actually conducted the work and cause the final draft of the paper to have cognitive implications different from those in the first draft.

It should be pointed out that Knorr-Cetina does not actually do this, because she pays little attention to how the last draft of the paper differed in specific cognitive content from the first.[3] She shows that negotiation was involved in the writing of the paper but fails to show what, if any, cognitive effect the negotiation had. But it would undoubtedly be possible to show that the negotiations over some papers would lead to different cognitive outcomes. I accept the fact that social variables and processes can have some influence on local knowledge outcomes, and, therefore, such demonstrations would not contradict my position.

The problem in such studies would be that because we truncated our analysis at the time of a local knowledge outcome's publication, we would not know the extent to which any of the choices or selections made in the local context made any real difference in the development of knowledge as accepted by the community. And when the evaluation phase progressed further to the time when the new paper was ignored, rejected as error, or accepted as fact we could not know whether the outcome might have been different had the negotiation leading to the form of the published paper had a different ending (for example, if the first draft of the paper had been published instead of the sixteenth).

The second research strategy which we could follow would be to study work which has been accepted into the core but which, while at the research frontier, involved controversy. Preferably we would have a situation such as that described by Pickering (1980, 1984), in which two or more solutions to a problem had been proposed, one of which won out and entered the core. Now we would know that the outcome did make a difference—that we were not simply studying the vast amount of noise generated in the attempt to construct a small number of facts. If it could then be shown how social variables influenced the form which each of the competing solutions took and that social processes influenced their evaluation, this evidence would support the constructivist position. We would, of course, have to avoid the pitfall of defining everything as "social" (including cognitive factors) and then arguing tautologically that the content of science is socially determined.

This type of analysis would still face serious problems. How would the sociologist characterize the variables to show relationships that would go beyond a type of low-level description? For example, we might find that in a particular case a certain type of social process or interaction contributed to a specific selection or decision. But given the scarcity of different solutions to the same problem it would be difficult to move beyond this level. What we could do is generate descriptions of the various social processes which, in general, influence the doing of science as opposed to specific cognitive outcomes. The constructivists have conducted much research which contributes to such an effort. Otherwise we would have to resort to using social variables so abstract and general (for example, "cognitive interests") that they would have little explanatory power.

A systematic empirical test of the basic theory of the constructivists could be conducted by performing a large-scale experiment. Suppose that we obtained the cooperation of five or more different chemistry laboratories to conduct an experiment. All the labs would be asked to solve the same clearly defined scientific problem. We would place observers, preferably using videotapes for systematic analysis, in each laboratory. Each scientist working on the project would be interviewed both before the experiment began and after. All interaction would be recorded and coded.

We would be interested in knowing the extent to which the solutions reached differed from each other. If a small amount of variance occurred in the content of the solutions, this would be strong evidence against the constructivist position. Scientists starting from a similar knowledge base and using rules of procedure which were similar from one setting to another would have come up with essentially the same answers. But if local contextual factors are as important as the constructivists say, we should see significant variation in the solutions reached by the different labs in this hypothetical experiment. If such variation were found we would try to link specific content differences with specific variations in the laboratories. Did the scientists indeed start from the same knowledge base (determined by interviews conducted before the experiment began)? Can any observed patterns of interaction be correlated with specific differences in outcome (based upon the taped interaction)? Did the availability of a specific type of equipment determine the outcome? It should be pointed out that even if these questions could be answered, the experiment would not tell us which factors determine which, if any, of the locally produced knowledge outcomes will become communal knowledge outcomes.

Because such an experiment would be expensive and time consuming it would probably be impossible to conduct; but it does suggest a research strategy which would be fruitful to pursue. In order to show how different contexts might contribute to the content of science produced, it would be theoretically useful to study different laboratories working on the same problem simultaneously. The difficulty of specifying a potential test of the basic hypothesis of the constructivists suggests that, in Lieberson's (1985) words, this research program formulated as a causal hypothesis might be "un-doable."

What is the future of the sociology of science? I believe that the sociology of science is too preoccupied by epistemological problems. In order to justify the sociological study of the cognitive content of the natural sciences, sociologists have concentrated their efforts on attacking realist orientations and supporting relativist ones. This work has been effective in helping to debunk the positivist mythology of science. Whether realism or relativism is "correct," however, cannot be proved by empirical evidence, research, or logical argument. The choice between these two epistemological positions is most certainly underdetermined and must be made on other bases.

Within the broad community of the social studies of science there appears to be a general consensus that science does not actually proceed according to the "rational" procedures depicted by the positivist philosophers. In the history and philosophy of science a new, more sophisticated type of realism seems to be emerging as the dominant view (Galison, 1987; Giere, 1988; Hull, 1988; Rudwick, 1985). In the sociology of science the dominant orientation is constructivism. Outside the community of the social studies of science, among sociologists in general, natural scientists, and the educated public, positivism probably remains the dominant position. Given that realism and relativism are in a sense incommensurable assumptions, it is very unlikely that further studies of the sort the constructivists have conducted will result in the overthrow of positivism in the larger community. It is much more likely that the studies done by the new realists will ultimately cause the larger community to adopt a more sophisticated view of science.

I hope that in the future we will see a greater integration of the constructivist approach with the more traditional concerns and methods of the sociology of science. It would be interesting and potentially very useful to see whether one could develop a nonrelativist version of constructivism. How much of the constructivist program could be maintained without the relativist epistemological position? Constructivists and "Mertonian" sociologists of science could fruitfully collaborate in exploring the many puzzles which have been posed by the constructivist approach. But as Kuhn ([1962] 1970) points out, pleas for cognitive "pluralism" are generally made by adherents of a paradigm which has "lost out." There is little likelihood that relativists and realists can put aside their epistemological differences and collaborate. Instead they will pursue their separate programs. It can only be hoped that constructivists will articulate more clearly their answers to some of the questions raised here, and that realists will develop empirical research designs which are capable of throwing light on some of the topics which emerge as central in the debate. These are the social processes through which consensus is achieved at the research frontier of science.

Appendix

Notes

References

Name Index

Subject Index

· · APPENDIX · ·

The COSPUP NSF Peer Review Experiment

Much of the data presented in this book is from the second phase of the NSF peer review study, which is described below.[1]

Design

The experiment was based on fifty proposals in each of three programs (chemical dynamics, economics, and solid-state physics), submitted and processed by the NSF in fiscal year 1976—the latest twenty-five proposals that had been funded by NSF and the latest twenty-five declined—for each program. (The actual funding percentage of NSF at the time was close to 50.) The sample did not include proposals for continuation of multiple-year funding, fellowships, and applications not processed entirely in this funding period.

We wanted to replicate as far as possible the NSF review procedure with independent reviewers not selected by the NSF. After providing copies of the fifty proposals in each field, NSF had nothing more to do with the experiment, which was conducted as follows.

1. For each program we chose a panel chairperson to select qualified reviewers of proposals.

2. Neither the chairpersons of reviewer-selection panels nor any of the participants knew which proposals had been funded and which declined, nor had they any knowledge of the NSF ratings or decisions.

3. Each panel chairperson selected 10 to 18 experts in the discipline to constitute a reviewer-selection panel, whose members acted, in effect, as surrogate program directors.[2]

4. The chairperson asked each prospective panelist whether he or she was willing to participate. There were virtually no refusals.

5. The number of reviewer selectors in each program was determined by the variety of proposal topics in the sample. Panelists were distinguished researchers in the substantive areas of the proposals. The chairperson determined which proposals were sent to which reviewer selectors.

6. Before the proposals were sent to reviewer selectors, we attempted to conceal the identity of their author(s). We removed title pages, lists of references, budgets, references in the text to the past work of the principal investigators, and identifying remarks or comments in the proposals. Such deletions are, of course, hardly a foolproof method for preventing identification, especially among experts in the same field. We asked reviewer-selection panelists whether they could identify the authors of proposals they received. A few could.

7. Each blinded proposal was sent to two reviewer-selector panelists, who were asked to list six scientists qualified to review a particular proposal. Some named as many as ten, others only three or four. The reviewer-selection panels never met. Their nominations for qualified reviewers were made, to the best of our knowledge, independently.

8. After eliminating from the lists of proposed reviewers anyone who had reviewed a particular proposal for NSF, anyone at the same institution as the principal investigator, and the principal investigator, we randomly assigned each reviewer a blinded or a nonblinded proposal.

9. Each proposal, blinded or nonblinded, was initially sent to four to seven reviewers. The number of reviewers for each proposal varied with the number nominated by the selection panel. The number of reviewers of different proposals also varies in the NSF process. When we had difficulty obtaining reviews, we asked the selectors to choose additional reviewers.

10. Reviewers were asked to evaluate nonblinded proposals using the criteria employed by the NSF. The letter discussing the criteria of evaluation was identical to the one used by the NSF.

11. Reviewers of blinded proposals were asked to identify authors of proposals if they could. Whether or not they attempted this, reviewers were instructed to evaluate the proposal strictly in terms of the content of the science contained in it. They were asked to ignore past track record, even if they thought they knew the principal investigator's identity.

Additional facts concerning the experiment should be mentioned.

a. All potential participants in the experiment were told that this was an experiment, that the actual funding decisions had already been made, and that their reviews would not affect the careers of the scientists involved.

b. In the course of designing the experiment, several decisions involving other possible strategies were made that could have affected its outcome. Some of these decisions closely parallel NSF decisions made in the selection of reviewers. For example, we decided to limit the number of proposals sent to any given reviewer to three, though some reviewers were named to review four or five proposals. (It has been NSF practice to use reviewers only a few times in any given year.)

c. As reported, any person named by a reviewer-selection panel who had reviewed the same proposal for the NSF was excluded. Some problems follow from this decision. Individuals nominated as reviewers by both the NSF and our panel might be viewed as most qualified, and excluding them could seem to leave only "residual" reviewers for our appraisals. But including them would have posed other problems. The correlations between the NSF ratings and the COSPUP ratings would be artificially increased by any overlap. For any NSF reviewer nominated by our panel, we did have the adjectival appraisals and the written comments from their NSF reviews, and could add these evaluations to our set of appraisals at a later stage in the analysis. When NSF reviewers who were selected by our reviewer-selection panel were added to COSPUP reviewers of nonblinded proposals, there was virtually no change in the average rating.

d. We did not have panel ratings in the experiment because we did not use panels, although they are used in some NSF programs. This limits our design in economics, in which the panel plays an important role in reaching NSF decisions (see S. Cole, Rubin, J. Cole, 1978, chap. 5).

Data on Principal Investigators

In addition to the peer reviews we obtained in the experiment, we collected data on the principal investigators of the proposals.[3] The following variables were included:

Citations to recent work.[4] All citations to work published between 1967 and 1976 were counted in the 1976 edition of the *Science Citation*

Index, except for anthropology and economics, for which we used the *Social Science Citation Index.* We also counted citations to co-authored papers published in these years on which the principal investigator was not the first author. (We were unable to do this for anthropology and economics). The program we used in analyzing these data had a default option which automatically coded all zeroes as equal to 1. The total number of citations received in this period was then converted to its log to base 10.

Citations to old work.[5] All citations to work published before 1967 were counted in the 1976 edition of the *Science Citation Index* or the *SSCI.* The program we used in analyzing these data had a default option which automatically coded all zeroes as equal to 1. This number was then converted to its log to base 10. For this measure we did not have citations to co-authored papers on which scientists were not first authors. However, the data for papers published in the last ten years, 1965 through 1974, showed that the total number of citations to first-authored and sole-authored papers and the total number of citations, including papers on which the authors were not first authors, were very highly correlated. In all eight fields for which we have data, the correlation was over .85. Therefore, the data on citations to work published prior to 1965 should adequately reflect the significance of older work.

Number of papers in last ten years.[6] The total number of papers published by the principal investigator between 1967 and 1976 was determined by making a count from the source index of the relevant science citation indices. The program we used in analyzing these data had a default option which automatically coded all zeroes as equal to 1. This number was then converted to its log to base 10. Because social science journals are not included in the *SCI,* and because the *Social Science Citation Index* began publication in 1972, we collected data on publications for social scientists from vitaes included in proposal jackets. For social scientists we constructed a productivity index which is described in S. Cole, Rubin, J. R. Cole (1978), app. B.

Recent NSF funding.[7] Past funding history was coded simply as the number of years since 1972 in which the principal investigator had received NSF funds. Thus the range on this variable was 0 to 5, with 5 indicating that the principal investigator had received NSF funds in all the five previous years, and 0 indicating that the principal investigator had not received any funds in the five previous years.

Rank of current department. The rank of current department was based upon data from the 1969 American Council on Education survey of graduate education. Principal investigators not employed in academic institutions were treated as having missing data; ranks of departments were converted so that a high score on this variable represents a high ranking. American departments not ranked in the 1969 study were placed at the bottom of the scale, at zero.

Type of current institution.[8] This is a dummy variable in which all principal investigators employed at Ph.D.-granting departments were given a score of 1 and all others a score of 0.

Rank of Ph.D. department. The rank of Ph.D. department was determined from the 1964 American Council on Education survey of graduate departments. Principal investigators who received their Ph.D. degrees from foreign institutions were treated as missing data for this variable. Ranks of departments were converted so that a high score on this variable represents a high ranking. American departments not ranked in the 1964 study were placed at the bottom of the scale, at zero.

Professional age.[9] The professional age variable is a function of the length of time since the principal investigator's Ph.D. degree was awarded. The variable was treated as a dichotomy, with those receiving their Ph.D. degrees in 1970 or before coded as 1, and those receiving Ph.D. degrees in 1971 or later coded as 0.

Academic rank. Full professors were coded as 3, associate professors as 2, and researchers, post-docs, assistant professors, lecturers, and instructors were all coded as 1. Principal investigators with no academic position were treated as having missing data for this variable.

Units of Analysis

Two basic units of analysis were used in the experiment. The first was the individual proposal. Ratings in each category were averaged, so for each proposal we had three separate mean ratings: those of NSF, those for the same version in the COSPUP experiment, and those for the blinded version.

The second unit of analysis was a pair consisting of the applicant and a rater. Again there were three sets of data: ratings by NSF reviewers, by COSPUP reviewers of nonblinded proposals, and by COSPUP reviewers of blinded proposals.

Response Rate

A total of 1,662 proposals (blinded and nonblinded) were sent out for review in the three disciplines. Our response rate was 72 percent; the NSF's rate for the same proposals was 78 percent. NSF obtained an 85 percent response rate in chemical dynamics, compared with 77 percent in the COSPUP experiment; the NSF figure was 77 percent in solid-state physics, compared with 72 percent in our experiment; and 69 percent compared with 67 percent in economics. The patterns of response to blinded and nonblinded proposals were also reasonably close. Evidently the blinding procedure did not much influence the level of response to requests for evaluation.

We sent an initial letter requesting cooperation with the experiment, and four follow-up letters over a period of five months to reviewers who had not responded. Finally, six months after the initial mailing, we telephoned all reviewers who had not returned the proposals.

Analysis of Variance

In order to determine the extent to which the reversals could be explained by bias or disagreement we used analysis of variance techniques. Because we did not want to make the usual statistical assumptions (such as normality) which must be made in a standard two-way analysis of variance, we used a components-of-variance model that did not require some of these assumptions but would be useful in answering the same substantive question.[10]

In order to assess the relative magnitude of contributions of the proposal evaluation method and the reviewer to the variation in ratings, we represent the rating y_{ijk}, given by the kth reviewer under method i to proposal j by

$$y_{ijk} = a_i + b_j + c_{ij} + e_{ijk},$$

where a_i is the overall average rating by evaluation method i ($i = 1$ for NSF and $i = 2$ for COSPUP), b_j is the differential effect of proposal j, c_{ij} measures the extent to which the rating on proposal j depends on the evaluation method, and e_{ijk} is the effect caused by the kth reviewer of proposal j by evaluation method i.

We consider a_i to be a fixed quantity and the remaining terms to be random with means equal to zero. Then we can decompose the variance associated with proposals under evaluation method i into three terms:

$$\text{Var}(Y_{ijk}) = \sigma_p^2 + \sigma_I^2 + \sigma_{R,i}^2$$

where $\sigma_p^2 = \text{Var}(b_j)$ reflects the intrinsic variability of the proposals: $\sigma_I^2 = \text{Var}(c_{ij})$ is the variability associated with the interaction of proposals and evaluation method; and $\sigma_{R,i}^2 = \text{Var}(e_{ijk})$ is the reviewer variance for method i.

If σ_p^2 is large relative to σ_I^2, $\sigma_{R,1}^2$, and $\sigma_{R,2}^2$, we interpret this to mean that it is relatively easy to distinguish the proposals independent of the evaluation method. However, if σ_I^2 is of the same order of magnitude as σ_p^2, this would suggest that dependence between proposal and evaluation method is masking some of the intrinsic proposal variability. As a consequence, the proposals would be ranked differently under the two evaluation methods. If, as actually occurs in these data, $\sigma_{R,1}^2$ and $\sigma_{R,2}^2$ dominate σ_I^2 and are of the same magnitude as σ_p^2 then reviewer variability will be so pronounced that two different evaluations will give dissimilar rank orders.

The estimates of σ_p^2, σ_I^2, $\sigma_{R,1}^2$ and $\sigma_{R,2}^2$ are presented in Tables A.1–3. The dependent variable for the analysis is the rating given the proposal by a reviewer. If we consider all the variance in an entire set of reviews (for example, all reviews done by both NSF and COSPUP nonblind reviewers for the fifty proposals), we want to know the sources of variance. There are four possible sources of variance, two of which turned out to be trivial in this study. Consider these four sources and the estimated effects for NSF and nonblind reviewers in solid-state physics (see Table A.1). The results for economics and chemical dynamics have parallel interpretations.

Table A.1 Components of variance of NSF and COSPUP nonblinded ratings

| Program | Proposal variance $\hat{\sigma}_p^2$ | Reviewer variance | | Interaction variance $\hat{\sigma}_I^2$ | Method difference $(\hat{a}_1 - \hat{a}_2)$ |
		NSF $\hat{\sigma}_{R,1}^2$	COSPUP $\hat{\sigma}_{R,2}^2$		
Chemical dynamics	23.67	55.91	56.67	1.18	2.73*
Economics	58.33	89.22	96.25	0.00†	2.14*
Solid-state physics	24.43	48.93	50.24	0.17	2.72*

Source: S. Cole, J. R. Cole, and Simon, 1981, p. 884, Table 3.
*NSF higher.
†The calculated estimate was negative.

Table A.2   Components of variance of NSF and COSPUP blinded ratings

| | Proposal variance $\hat{\sigma}_p^2$ | Reviewer variance | | Interaction variance $\hat{\sigma}_I^2$ | Method difference $(\hat{a}_1 - \hat{a}_2)$ |
| | | NSF $\hat{\sigma}_{R,1}^2$ | COSPUP $\hat{\sigma}_{R,2}^2$ | | |
Program					
Chemical dynamics	18.38	55.91	83.10	2.59	2.59*
Economics	52.75	89.22	100.49	3.76	4.85*
Solid-state physics	27.46	48.93	72.14	3.17	3.02*

*NSF higher.

Table A.3   Components of variance of COSPUP nonblinded and blinded ratings

| | Proposal variance $\hat{\sigma}_p^2$ | Reviewer variance | | Interaction variance $\hat{\sigma}_I^2$ | Method difference $(\hat{a}_1 - \hat{a}_2)$ |
| | | COSPUP nonblinded $\hat{\sigma}_{R,1}^2$ | COSPUP blinded $\hat{\sigma}_{R,2}^2$ | | |
Program					
Chemical dynamics	23.91	56.67	83.10	0.00†	.15*
Economics	54.41	96.25	100.49	1.05	−2.71
Solid-state physics	28.11	50.24	72.14	2.90	−0.31

*Nonblinded higher.
†The calculated estimate was negative.

First, reviewers' responses to proposals differ because proposals differ in quality. This variance is easily dealt with statistically by taking as a rough indicator of the quality of a proposal the mean of all its ratings by both NSF and COSPUP nonblind reviewers. This leads to a measure of the variation in quality of proposals (σ_p^2, above) that can be compared with other sources of variation. The estimated proposal variance for the solid-state physics proposals was 24.43.

Second, the NSF review procedures and the COSPUP procedures were not identical. On the average, there may be systematic differences between NSF reviewer responses to all proposals and COSPUP reviewer responses. This "method effect" can be observed in the differences in the mean ratings of proposals by NSF and COSPUP reviewers. As noted above, the COSPUP reviewers were on average slightly harsher than NSF reviewers. In the NSF–COSPUP nonblinded comparison for solid-state physics the estimated overall difference is 2.72 points, with NSF higher. Because funding decisions are based on rankings, this method effect is not important (but it was not ignored in the mathematical analysis).

Even after we compensate for the average methods effect, we find that reviewers may disagree in their ratings of a proposal because they are members of two groups selected differently—NSF reviewers as opposed to COSPUP reviewers. This "interaction" effect (σ_I^2, above) between proposals and evaluation method is important. It is the key component in estimating whether there appears to be any systematic bias among NSF program directors in the selection of reviewers. If bias existed in the selection of NSF reviewers, or if the two groups of reviewers had significant differences in the way in which they evaluated the proposals for *any* reason, we would expect the interaction effect to be large. If it is large, then the NSF reviewer group and the COSPUP reviewer group evaluated proposals differently. If it is small, they did not and we would not be able to detect any bias in the selection of the NSF reviewers. It turns out that the estimated interaction σ_I^2 is trifling for each of the three fields, so there is no evidence of disagreement between the two selection methods aside from apparent disagreement resulting from the reviewer variability.

Finally, variation that remains is denoted by $\sigma_{R,i}^2$, above, and measures the reviewer variation within a given evaluation method i. The reviewer variances were estimated to be 48.93 and 50.24 for solid-state physics. These numbers are rather larger than the estimated proposal variance of 24.43. Thus the reviewer brings to this process a higher variance than does the proposal. Of course, the average of several reviewers will have a lower variance; indeed, the average of four reviewers will have a variance of $48.93/4 = 12.23$ (NSF) or $50.24/4 = 12.56$ (COSPUP), but these are still not tiny compared to the proposal variance. This fact explains why the data exhibit so many reversals; they reflect substantial reviewer variance and not any fundamental disagreement between NSF and COSPUP reviewing methods or substantive evaluations.

To explain the reversals, then, we must look at two sources of variance: differences among the proposals and differences among the reviewers of a given proposal (Tables A.1–3). In the two physical sciences the variance among NSF reviewers and COSPUP nonblind reviewers of the same proposal is approximately twice as large as the variance among the proposal means; in economics the reviewer variances are about 50 percent larger than the proposal variance. If the pooled proposal mean (the mean of both sets of ratings in each comparison) is taken as a rough indicator of the quality of the application, we can see that the variation in quality among the fifty proposals is small compared with the variation in ratings among reviewers of the same proposal. We have treated the reviewer variances as rough indicators of disagreement among reviewers. In all three fields there is a substantial amount of such disagreement. It is the combination of relatively small differences in proposal means and relatively large reviewer variation that creates the conditions for reversals.

Thus far I have confined the discussion of the results of the experiment to a comparison of the reviews obtained by the NSF and by the COSPUP nonblinded reviewers. We also conducted a components-of-variance analysis in which the blinded and nonblinded COSPUP reviews were compared (see Table A.3). The amount of variance explained by the difference in proposal means was about the same when we considered the blinded reviews as when we considered the two nonblinded ratings. The method components show that in economics and solid-state physics there is a slight tendency for ratings of blinded proposals to average lower than the ratings of the same proposals in their original form. Reviewer disagreement regarding blinded proposals was a bit larger than when the proposals were judged in their original form. For COSPUP reviewers in solid-state physics, for example, the reviewer variance for blinded proposals was 72.14, compared with 50.24 for nonblinded proposals. The estimates of the interaction effect were all small. Of particular interest is the interaction effect when the two COSPUP ratings are compared. For these two sets of ratings, reviewer selection was controlled in the same manner and the interaction can therefore be strictly attributed to blinding and not to reviewer differences. The variances representing this interaction effect are 0.0, 1.05, and 2.90 in chemical dynamics, economics, and solid-state physics, respectively. These data lead to the conclusion that it is reviewer disagreement, rather than blinding, which accounts for nearly all the reversals in a comparison of blind and nonblind reviews. These data reinforce the conclusions that characteristics of the applicant had very little to do with how their proposals were evaluated and that reversals were a result of cognitive disagreement among members of a research area.

Notes

1. Nature and the Content of Science

1. *Science, Technology, and Society in Seventeenth-Century England* continues to be the focus of research and debate in the history of science. See, for example, Cohen (1990) and particularly Merton's "Afterword." On the foci of attention aspect of the monograph, see Zuckerman (1989a).
2. Woolgar (1988) has called this conclusion "Mannheim's Mistake."
3. Undoubtedly the emergence of a relativist sociology of science was affected not only by specific developments in the history and philosophy of science but by the general climate, which favored antirealist, relativist views. We can see similar approaches developing in other fields, such as deconstructionism in literary theory.
4. Kuhn uses the term "paradigm" in many different ways. See Masterman (1970). A good description of Kuhn's usage is provided by Stewart (1990): "A paradigm is a constellation of beliefs, values, procedures, and past scientific achievements that is shared within a community of scientists, directs their research activity, and is learned during their training or in their common research experiences" (p. 4). There is a common misperception that "normal" science means unimportant or trivial science. This is incorrect. Virtually all the work done in physics since the development of relativity and quantum mechanics, for example, including the work for which Nobel Prizes have been awarded, is normal science.
5. Brush (1974), in his provocative article entitled "Should the History of Science Be Rated X?," provides additional evidence on this point from his research on the caloric theory and the experiments of Joule. Most textbooks on physics indicate that it was Joule's experiments, conducted in the middle of the nineteenth century, which demonstrated the equivalence of heat and mechanical energy. These experiments supposedly overthrew the earlier caloric theory. But Brush points out that, as a result of other work that had been done, most people no longer believed in

the caloric theory by the time Joule conducted his experiments. The textbook version of the role of Joule's experiments is thus merely a historical reconstruction, and not what really happened.

6. Laudan (1990) discusses this issue in detail.

7. Strictly speaking, a paradigm cannot be both incompatible and incommensurable. If two paradigms are incommensurable this means they cannot be brought to bear upon each other and therefore cannot be incompatible. Paradigms which are incommensurable look at the world so differently that a scientist schooled in one has great difficulty even understanding the other.

8. Woolgar (1988) has pointed out that some constructivists are ambivalent about what he calls "essentialism," the view that the empirical world imposes constraints on the development of cognitive knowledge. For example, Barnes (1974) and at least the earlier Mulkay (1979) admit that the external world constrains the development of knowledge. Woolgar criticizes these constructivists for their failure to maintain a pure relativism. To the extent that constructivists are willing to accept that the empirical world constrains the content of science and that the extent of this constraint is an empirical question, there is little or no difference between their position and that taken in this book.

9. There is considerable disagreement on the validity of Collins' conclusion that replication is rarely done. This controversy will be discussed in Chapter 2.

10. Excellent studies which present similar views of the influence of random and haphazard processes on the doing of scientific research are Gilbert and Mulkay (1984) and Knorr-Cetina (1981).

11. Laudan (1977) presents an argument that science is "rational," but he defines this concept in a pragmatic way which does not correspond to the sense in which the term is used here.

12. This point has also been made by Elkana (1981), p. 37, and Hull (1988), p. 7.

13. Most case studies of research conducted at the frontier provide ample anecdotal evidence on the low level of cognitive consensus. A good example is the study by Hull (1988) which illustrates the high level of disagreement among members of different "schools" and among members within the same school. This disagreement is frequently based upon social as well as cognitive grounds.

14. Winstanley (1976) found the Watson and Crick model to be incorporated almost immediately into research-level texts, although it took longer for the new knowledge to enter school texts and popular literature.

15. It has been claimed that a cure for AIDS developed in Kenya has been ignored as a result of racial prejudice. This is indeed possible, though highly unlikely given the potential profit to be made from finding such a cure. What would not be possible would be the adoption, on the basis of racial prejudice or any other reason, of an ineffective "cure."

16. A similar argument is strongly made in the analysis by Hull (1988) of the development of two biological specialties: numerical taxonomy and cladistics.

17. The research I have done on consensus "black boxes" the cognitive content of the science. I do not study how social processes or variables influence any particular scientific ideas. This is not a problem for my work, because I make no claim to showing that the specific content of any science is socially determined. In fact, as my analysis in Chapter 2 will make clear, I remain skeptical as to whether it is possible to perform such sociological studies.

18. My position is similar to that of the philosopher of science Ronald N. Giere (1988), who has made similar critiques of the social constructivists. He develops a four-cell typology in which he classifies orientation toward science according to whether it sees scientific representation as realist or antirealist and whether it views the judgments made by scientists as rational or natural: "I will use the label rationalism for the view that there are rational principles for the evaluation of theories. I realize that this terminology has the misleading consequence that many 'empiricists' turn out to be 'rationalists,' but I can think of no better term. Naturalism, by contrast, is the view that theories come to be accepted (or not) through a natural process involving both individual judgment and social interaction. No appeal to supposed rational principles of theory choice is involved" (p. 7). Traditional positivists fall into the realist-rational cell; constructivists into the antirealist-natural cell. He classifies himself in the realist-natural cell. In so far as I understand his argument, I would also see the position taken in this book as falling into that cell.

2. Constructivist Problems in Accounting for Consensus

1. Some of the sociologists whom I include in my discussion of constructivism do not accept this label for their own work. The divisions among the constructivists are of more interest to those working in the social studies of science than to those who are not. As I shall point out later in this chapter, I believe that there is an underlying core which holds together the work of most of these sociologists.

2. I disagree with Gieryn (1982) on this point. Although many anticipations of current concerns can be found in Merton's work, the constructivists focus on problems that were only peripheral to Merton's concerns. Gieryn has since changed his orientation and is now supportive of the constructivist approach to science studies.

3. This work was done by Guillemin's group at the Salk Institute before the period of Latour's participation.

4. The reader will note that the word "discovered" is in quotes here. Social constructivists do not like to use this word, because it implies that a scientific fact or theory has been in existence throughout time and has only been discovered by the scientist. They argue that because facts or theories only exist as a result of the scientists' behavior, they are not discovered but socially constructed. I omit quotes in all subsequent uses, but the reader should keep in mind that the concept of discovery has no place in the work of constructivists.

5. The first principle of the "strong program" in the sociology of knowledge

as laid out by Bloor (1976) is that of "causality," that is, the sociologist must try to identify the social variables "which bring about belief or states of knowledge" (p. 4).

6. This research program has had a great deal of appeal to the ethnomethodologists. Believing that if "reality" in the natural sciences could be shown to be socially constructed, this would make a strong case for the social construction of reality in social life, the ethnomethodologists have adopted science as a research site. Harold Garfinkel and some of his students have published ethnomethodological studies of science. See, for example, Garfinkel, Lynch, and Livingston (1981) and Lynch (1985).

7. A similar critique is made in Zuckerman (1988).

8. Much of the criticism made by the constructivists of the "Mertonian" school of the sociology of science has focused on Merton's work on the norms of science ([1942] 1957a). When I use the term "norms" in this context, I do not use it to refer to Mertonian norms; rather I use it to index the socially generated and maintained beliefs which exist in the scientific community at any given time and which influence the way in which science is done (Laudan, 1977). A good portion of these norms are cognitive norms or beliefs (Mulkay, 1977, p. 246; Stehr, 1978). For example, in sociology the belief that certain types of data should be analyzed using log-linear models or structural equations is a current methodological norm. The Mertonian norm of universalism is discussed in Chapters 7 and 8.

9. It is, of course, possible and desirable to do micro-level studies of how scientists react to the work of others. But most of the studies published to date have concentrated more on the creation of new work in the laboratory and paid little attention to observing how scientists evaluate the work of others.

10. For a similar analysis of the Gargamelle experiment see the review by Gingras and Schweber (1986) of Pickering (1984).

11. For a full and more convincing development of an evolutionary theory of selection in science see Hull (1988).

12. Other studies which have conducted content analyses of scientific papers in order to understand the function of citations support this view (Adatto and S. Cole, 1981; Chubin and Moitra, 1975; S. Cole, 1975; Moravcsik and Murugesan, 1975).

13. What Latour means by a "black box" in this context is a piece of scientific equipment or a procedure which is accepted by the community as yielding valid results without questioning any of the assumptions that are necessary in utilizing the equipment or conducting the procedure. Latour and other constructivists argue that the dependence of scientists upon black boxes and complex inscription devices shows that rather than empirical evidence being the basis for new facts, it is the assumptions which lie behind the social construction of the black boxes. Galison (1987) conducts an analysis of how certain procedures become black boxes: "While it is *logically* possible to challenge any aspect of an experimental procedure, certain techniques eventually are well enough understood to leave

the zone of disputed assertions. Thus in 1932 it was still plausible to argue that all nominally high-energy tracks were being straightened out by turbulence in the gas. By 1936 Anderson's cloud-chamber technique had improved to the point where such challenges would have seemed ludicrous" (p. 130).

14. Some of the positivist philosophers, such as Nagel (1961), made use of a similar concept.

15. Note here Watson's account that X-ray evidence was forcing Rosalind Franklin to abandon her own position and accept the view that DNA was a helical structure. This is another of the multitude of examples of how scientists change their views when presented with empirical evidence they cannot explain in other ways.

16. The hypothesis that the reaction of the scientific community to a new discovery is influenced by the social characteristics of its author(s) will be analyzed in Chapters 6 and 7.

17. Although selective quotation from Merton's large body of writing on science can make him seem to be a "naive" positivist, this was far from true. Consider his 1937 discussion of how facts are determined: "There are a number of other ways in which social interaction influences the development of science. The laws used in science are selected from a number of possible laws which adequately state uniformities between the observed facts. *Hence, the selection can not be made solely upon the basis of sheer correspondence to the facts.* In other words, in the selection of a law which is stated to be 'true,' its truth (correspondence to the facts) is a necessary, but not sufficient, condition of its selection. *The law is ultimately found acceptable because it fits into a theoretical structure the form of which is determined by preconceived ideas of what a theory should be* . . . In short, from among the various theories which satisfy the observations the one chosen is selected because of the intellectual satisfaction which it affords. *But the criteria of what constitutes intellectual satisfaction in any given period arise largely from the cultural scheme of orientation*" (1937, p. 169, italics added).

18. It is interesting to point out that the successful strategy of Guillemin involved the development of a method which enabled greater precision in determining what the chemical structure of a substance was. This greater precision made some results more believable and others less so. This example is similar to that used by Latour (1987) of the more powerful telescope which allowed closure of a scientific controversy. Precise measurement seems to be one characteristic of the content of science which is valued in many fields.

3. Constructivist Problems in Demonstrating Causality

1. In this case research has shown that the increased use of Cesarean sections cannot be explained by medical reasons. See Summey (1987). It would, of course, be an interesting sociological problem to examine how social factors influence the evaluation and interpretation of data on infant mortality and birth defects.

2. Indeed, philosophers such as Laudan (1977) have argued that sociological explanations should be sought only after an exhaustive search for cognitive explanations.

3. In more recent work Pickering (1989) somewhat grudgingly acknowledges that the results of experiments sometimes do cause scientists to change their minds.

4. This analysis is a good example of theory-laden sociological "observations."

5. This would appear to be another example of the confusion between cognitive and social influences on science. In this part of her monograph Knorr-Cetina seems to be primarily interested in the extent or lack of extent to which *cognitive* rules govern the decisions and behavior of scientists in the laboratory, but in this example she seems to be referring to a social rule. Who had the right to use the lab under what circumstances is a rule associated with the particular organization in which the research was conducted.

6. Collins (1975, p. 221) makes a similar critique of the strong program.

7. Although the reasons for decisions made in the laboratory were omitted from the final draft of the paper, it appears that this draft actually contained more methodological information than the first. Thus the final draft was probably more useful to anyone wanting to replicate the work than the first would have been.

8. This point is also made by Zuckerman (1988), p. 555.

9. On the significance of published science see also Garber (1989).

10. Other examples of work which is descriptively interesting but does not contain any discussion of sociological variables are Chubin and Moitra (1975); Crane (1980); Moravcsik and Murugesan (1975); and Small and Griffith (1974).

11. Mitroff strangely attributes such views to Merton, who has long emphasized the emotional commitment of scientists and their ambivalence to the norms of science. See Merton (1976), pp. 56–64.

4. Luck and Getting an NSF Grant

1. Sections of this chapter appeared in an earlier form in S. Cole, J. R. Cole, and Simon (1981) and J. R. Cole and S. Cole (1981, 1985). I thank Jonathan R. Cole and Gary Simon for their help with these sections. The research on the peer review system reported on here was sponsored by the Committee on Science and Public Policy, National Academy of Science and supported by a contract with the National Science Foundation. Although some (Chubin and Hackett, 1990) have classified this study as being "agency sponsored," this designation is misleading. The NSF had no control over this research. The contract between the NSF and NAS specifically stated that the researchers would have complete freedom to collect needed data and if that freedom were abridged the contract would be cancelled. The NSF made no attempt to influence the outcome of the study and was extremely cooperative in providing access to all necessary data.

2. My collaborators on this project were Jonathan R. Cole and Leonard Rubin. Some of the results of this research have been previously published in S. Cole, Rubin, and J. R. Cole (1978); J. R. Cole and S. Cole (1979, 1981, 1985); S. Cole, J. R. Cole, and Simon (1981).

3. Another major form of peer review is that utilized by the National Institutes of Health. At NIH peer review is done primarily by a panel of expert scientists, a study section, which reviews the proposals and then meets to evaluate them. This form of peer review was not studied. The results reported in this chapter apply only to the basic science programs at NSF and may not be generalized to any other form of peer review.

4. For a detailed discussion of the factors taken into consideration in the design of the sample, see S. Cole, Rubin, and J. Cole (1978), app. B.

5. For a more complete description of the procedures and methods used in this experiment see the Appendix.

6. The reviewer selectors were sent a blinded copy of the proposal.

7. Some have argued that the highly specialized state of modern science does not permit more than a dozen or so scientists to be capable of reviewing any given proposal. The COSPUP experiment enabled us to test this hypothesis. If the number of eligible reviewers were, in fact, small, we would expect that a fairly high proportion of the original NSF reviewers would also have been selected by the experimental reviewer selectors. In each of the three programs, about 80 percent of the NSF's reviewers were not selected by either of the two COSPUP selectors, about 15 percent were selected by one of them, and about 5 percent were selected by both. These data suggest that the pool of eligible reviewers for most proposals is at least 10, and given the low overlap rates we found, I would predict that if other equally qualified selectors were employed we would find it to be substantially larger than 25. The size of the pool (N) can be estimated as follows. This approximation assumes that each COSPUP selector makes 6 equiprobable choices from the pool of N. Suppose that individual A had been selected by NSF. Then P (the probability that A will be selected by COSPUP selector) $= 6/N$. Let $p = 6/N$. Then P (no overlap) $= (1 - p)^2$; P (one overlap) $= 2p(1 - p)$; P (double overlap) $= p^2$. The proportions $(1 - p)^2 : 2p(1 - p) : p^2$ correspond closely to the proportions observed when $p = 0.1$. This suggests that $N = 60$. The approximation is not perfect, of course. The field of economics produces a surprising number of double overlaps. In actual practice, of course, there is not always a clear distinction between eligibles and noneligibles, and the numbers vary according to subspecialties.

8. NSF may decide to fund a piece of the proposed scientific work and reduce the amount of the grant accordingly. Here I do not differentiate such a grant from a full grant.

9. The Phase I data showed a very high correlation between the mean rating of the reviewers and the program director's decision. The mean rating generally explained more than 75 percent of the variance on decision. For the proposals used in the experiment there was also a very high correlation between the mean rating of NSF-selected reviewers and the decision of the NSF on whether to fund or decline the proposal.

10. We classified reversals within quintiles as follows. The proposals were grouped into quintiles based on COSPUP rank, the first (best) quintile containing proposals with ranks 1, 2 . . . 10. A proposal was counted as reversed if it was in the upper twenty-five by one set of ratings and in the lower twenty-five by the other set. When there were ties in the mean ratings crossing quintile boundaries, proposals were apportioned among the categories involved. This rule may result in a noninteger number of reversals.

11. See Chapter 5 for a discussion of field differences in consensus.

12. If one takes two independent observations from a normal population with standard deviation σ, the expected absolute difference is $2\sigma/\sqrt{\pi}$. This statement is a reasonable approximation even when the population values do not follow a normal distribution. The numbers in the text reflect a reviewer standard deviation of 7.49 and a proposal standard deviation of 4.36. Note that $(2/\sqrt{\pi})7.49 = 8.45$ and $(2/\sqrt{\pi})(4.36^2 + 7.49^2)^{1/2} = 9.78$. The value $(4.36^2 + 7.49^2)^{1/2}$ reflects the fact that a randomly selected review incorporates both the proposal standard deviation and the reviewer standard deviation.

13. An additional 70 never reached reviewers because of address changes.

14. There was some variability among the three fields: 82 percent of the reviewers in chemical dynamics answered the questionnaire on reputations, 86 percent in solid-state physics, and 74 percent in economics.

15. Professor William Kruskal, letter to Dr. John Slaughter, director, National Science Foundation, February 4, 1982.

5. Consensus in the Natural and Social Sciences

1. Parts of this chapter appeared in an earlier form in S. Cole (1983), a paper originally written while I was a fellow at the Center for Advanced Studies in the Behavioral Sciences in 1978–79. Earlier versions were presented at the Spring Institute of the Society for Social Research, Department of Sociology, University of Chicago, May 5, 1979, and at the annual meeting of the Sociological Research Association, August 27, 1980, New York City.

2. Strategic research sites for studying the evaluation of science are the grant review process, the journal review process, and the evaluation of individual scientists for receipt of rewards. The last is discussed at length in Chapters 7 and 8.

3. As I pointed out in Chapter 2, constructivists such as Latour deny that these attributes have any influence on the evaluation of new contributions.

4. For Comte, of course, unlike his followers, sociology was at the apex of the hierarchy rather than at the bottom.

5. Before adopting the constructivist orientation, Knorr-Cetina, like most others, assumed that there were higher levels of consensus in the natural than in the social sciences. In her 1975 paper she seeks to explain why these differences exist.

6. For a reply to Hargens see S. Cole, Simon, and J. R. Cole (1988).
7. Even Meyer's data somewhat understates the number of pages published by these two journals, because he does not count the small number of articles in which no citations were made.
8. The space in the *AJS* devoted to book reviews has been excluded.
9. The Polish journal does not use blind refereeing (neither do U.S. physics journals). The articles are refereed by members of the editorial advisory board. Also, like U.S. physics journals, the Polish journal probably does not have as much competition for space as do U.S. sociology journals.
10. The conclusion that science was universalistic was oversimplified. For a reexamination of this problem see Chapters 7 and 8 and J. R. Cole (1987).
11. Details on the sample and measurement of key variables are presented in S. Cole (1978).
12. For an analysis of how social processes produce the high correlation between the quality and quantity of published research see S. Cole (1978).
13. After standardization, the data from all five fields were combined. For each dependent variable in the model two different regression equations were computed. The first equation entered all of the independent variables plus a dummy variable for four of the five fields. The second equation entered all the independent variables plus the dummy variables plus a series of interactive terms designed to see if the different independent variables had different effects in the different fields. The differences in the amount of variance explained by the equation with the interaction terms and the equation without them were not statistically significant at the .05 level.
14. For a review of the literature on this topic see Simonton (1988).
15. For more details on how these data were collected and key variables measured see S. Cole (1979).
16. See S. Cole (1978) for a description of the survey.
17. The data for the *Physical Review* were provided by Cullen Inman of the American Institute of Physics.
18. If a contribution was made prior to receipt of the doctorate, it was counted as a negative number in computing the averages. Only five of the physicists and one of the sociologists published a five-citation discovery prior to receiving the doctorate.
19. The first two sets of studies reported in this chapter do not, of course, contain any direct measures of consensus. They instead hypothesize what effects disciplinary differences in consensus should have if they do exist and then look for these differences. Others have used a similar logic. The power of such studies depends upon the extent to which the reported correlations are believable consequences of consensus and *not* believable consequences of other variables. For example, Hargens (1975) shows a correlation between journal rejection rates and the average length of doctoral dissertations. Hargens and Hagstrom (1982) argue: "It is likely that relatively long dissertations are the effect of low consensus, since more evidence and argument is necessary to justify conclusions; and the

effect of low codification, since theories and procedures cannot be stated so succinctly" (p. 189).

20. Responses obtained from such a questionnaire item have several evident limitations. One is that the question asks respondents to evaluate the significance of a colleague's work without taking into account the extent to which they are actually familiar with that work; some of the scientists answering the question may have only a hazy superficial idea of a rated scientist's work, and others may have detailed familiarity with it. We do not know whether the degree and extent of this type of knowledge varies from field to field.

21. For a more complete discussion of this analysis see S. Cole, J. R. Cole, and Dietrich (1978). Although the Gini coefficient has been used as a measure of consensus in science, there is a significant conceptual problem in its use. The Gini coefficient simultaneously measures two concepts. One is consensus, and the other is the dispersion of recognition in a particular field. It would, for example, be possible to have complete consensus in a field about the importance of a group of scientists' work, and yet all the scientists could be assessed as being equally important, that is, receiving the same number of citations. Thus, if one has 100 scientists rating the work of 10 other scientists, all 100 raters might agree that the work of each of the 10 rated scientists was equivalent in value. Under such a circumstance the Gini coefficient would be zero but consensus would be complete. If the 100 raters had complete consensus but decided that there were wide differences in the significance of the contributions of the 10 rated scientists, the Gini coefficient would be high. In practice it is impossible to distinguish the extent to which a Gini coefficient is influenced by the dispersion of evaluations of particular works or the level of consensus among scientists on the significance of work. In defense of the use of the Gini coefficient in comparative studies such as this one, it can be claimed that there is now significant evidence that the distribution of recognition in all scientific fields is highly skewed. If dispersion of recognition is approximately the same in all scientific fields, differences in the Gini coefficient should measure differences in consensus. Because of the difficulty of actually proving this, however, studies using the Gini coefficient should be seen as suggestive, not conclusive.

22. I thank James S. Coleman and Judith Tanur for their suggestions on summaries and statistical analyses of these data.

23. Hargens (1988b) has criticized the statistical adequacy of prior analyses of the peer review data. He points out that most statistical measures used in the past to assess the extent to which referees agree are inadequate *comparative* measures of consensus, because they are influenced by the heterogeneity of items being evaluated (research proposals in this case) as well as by the consensus of reviewers on the merits of each item. Specifically, Hargens argues that the figures reported in Tables 4.5 and 4.6 are influenced by the greater heterogeneity of proposals and articles in the social sciences and that lower levels of consensus in these fields will thus yield equivalent ratios. Hargens' statistical point is certainly cor-

rect, but data presented in Table 5.7 show the average standard deviation of ratings within proposals, a statistic that is not a ratio and therefore not influenced by field differences in proposal heterogeneity. This statistic is a valid measure of the extent to which scientists agree and is appropriate for making cross-field comparisons. These data indicate that for the Phase I data economics has more cognitive consensus than some of the natural sciences and that anthropology has cognitive consensus at the same level as biochemistry and geophysics. Hargens (1988b) claims that this figure is depressed for fields such as economics, which had more cases in which there was only one ad hoc mail reviewer of a proposal, thus making the standard deviation zero for that proposal. But even when all such cases are excluded from the analysis there is no change in the substantive conclusions reached from Table 5.7. This conclusion is also supported by the results presented in Table 5.4, where the measure of consensus is not influenced by differences in heterogeneity and where each scientist had a large number of raters. In Cicchetti's (1991) secondary analysis of the Phase II data he excluded the economics cases which had only one reviewer and still found higher levels of agreement in economics than in the two natural sciences.

24. Just as sociologists have assumed consensus differences between the natural and the social sciences, they have also assumed codification differences. Zuckerman and Merton (1973a) use the "immediacy effect," or proportion of references to recent work, as an indicator of codification. Elsewhere (S. Cole, 1983) I show that there are no systematic differences in immediacy effects between the natural and the social sciences. I do find, however, that all sciences have higher immediacy effects than does a sample of papers from English literature.

25. Even if the subject matter was used as the unit of analysis, empirical research would be required to see whether the potential for disagreement is greater for topics in the social sciences than for those in the natural sciences.

6. Evaluation and the Characteristics of Scientists

1. Sections of this chapter appeared in an earlier form in S. Cole, Rubin, and J. R. Cole (1978). My 1978 co-authors are not, of course, responsible for the interpretation of the data made in this chapter.

2. For a further development of the theory see Merton (1988).

3. Other research investigating the problem of accumulative advantage will be discussed in detail in Chapter 7.

4. See S. Cole, Rubin, and J. R. Cole (1978), app. B, for a discussion of the factors taken into consideration in the design of the sample.

5. See the Appendix for a description of how all the variables were measured.

6. It should be pointed out that some of the scientists who, in these data, appear to have poor track records are young scientists who have not had the opportunity to demonstrate their competence.

7. See the Appendix for a description of how this and the other independent variables were measured.

8. Because the mean and standard deviations on the relevant variables differ significantly from program to program, it was necessary first to standardize separately all the data within each field before combining data on applicants from different programs. For example, the number of citations received by a biochemist will not be expressed in absolute number but rather in terms of the number of standard deviations above or below the mean for biochemists. Thus a biochemist who is one standard deviation above the mean for biochemistry will be treated as equivalent to an anthropologist who is one standard deviation above the mean for anthropology, despite the fact that the biochemist has many more citations than the anthropologist. I have converted absolute scores on the variables into scores relative to other individuals in the same program. These standardized scores are comparable across programs.

9. It is worth noting that there are generally high correlations between the number of papers a scientist has published and the number of times she has been cited. In fact, for biochemistry, the field that on the average shows the highest correlation between the three productivity variables and the ratings, I found that all three variables together explained only 17 percent of the variance, only 1 percent more than was explained by citations to recent work alone.

10. I have referred to the amount of variance explained by these applicant characteristics as being "low," but we should keep in mind that sociological explanations frequently fail to explain large portions of the variance on social outcomes. The large body of sophisticated research done on status attainment, for example, finds that only about 25 percent of the variance on income in the United States can be explained (Jencks, 1972).

11. For this analysis of the influence of applicant characteristics on communal evaluation, I first replicated the Phase I analysis for the three programs (chemical dynamics, economics, solid-state physics) included in Phase II. Again I used the applicant-reviewer pair as the unit of analysis. The results were very similar to the Phase I results reported in Table 6.1. The eight characteristics of applicants explained 15 percent of the variance in chemical dynamics (16 percent in Phase I), 18 percent of the variance in economics (21 percent in Phase I), and 14 percent in solid-state physics (17 percent in Phase I). The Phase II data used only eight characteristics of the applicants. The dummy variable indicating whether the applicant was employed in a Ph.D.-granting department was excluded.

12. When such analyses were conducted they showed high multiple R's but substantially lower adjusted multiple R's.

13. In the Phase I study we found that the same set of applicant characteristics explained between 17 percent of the variance and 70 percent of the variance in decision. In most of the ten programs, about 35 percent of the variance in decision was explained by applicant characteristics (S. Cole, Rubin, J. R. Cole, 1978, chap. 5). For this analysis we used a probit regression technique, because the dependent variable was a dichotomy.

14. See the Appendix for a description of the "blinding" process.
15. In chemical dynamics 50 percent of reviewers made no guess and 10 percent made an incorrect guess; in economics 63 percent made no guess and 2 percent an incorrect guess; in solid-state physics 49 percent made no guess and 7 percent an incorrect guess. In some cases, incorrect identifications included past collaborators of the actual principal investigators, or persons doing very similar work. In general, reviewers were somewhat more likely to identify the proposals of more eminent applicants. For these data and their analysis see J. R. Cole and S. Cole (1981), pp. 45–49.
16. When I combined the data from the three programs I first standardized each variable separately within field. I then combined the data on the standardized scores. It turned out that the only variables which required standardization were the productivity and citation measures. The overall results using standardized and nonstandardized data were virtually identical. I report results using standardized data for the productivity variables and nonstandardized data for the other variables.
17. For the data and a more detailed discussion see S. Cole, Rubin, and J. R. Cole (1978), pp. 35–43.
18. These data are described more fully in S. Cole (1978).

7. Is Science Universalistic?

1. This chapter began as a collaboration with my brother, Jonathan R. Cole. His paper on the same topic has been published in Jon Elster, ed., *Justice and the Lottery* (Cambridge: Cambridge University Press, 1987).
2. If universalism meant that rewards should be distributed on the basis of scientific role performance, in order to study universalism in science it would be necessary to have a measure of that role performance. The advent of the *Science Citation Index* provided sociologists of science with a relatively simple quantitative tool for measuring the impact of scientific work. For a discussion of the problems involved in using this tool see J. R. Cole and S. Cole (1973), chap. 2, and S. Cole (1989). For a critique of the use of citations as indicators in the sociology of science see Edge (1979).
3. Merton first introduced this concept in his lectures on sociological theory at Columbia University.
4. For the same type of analysis and logic based upon a sample of Israeli scientists and engineers see also Shenhav and Haberfeld (1988).
5. For a review of more recent literature on gender and science, see Zuckerman (1991). There is now substantial evidence that male scientists on average publish about twice as much as female scientists and that the various explanations of this productivity gap which have been offered (such as family obligations or women not having as much access to research universities) have all failed to "wash out" the effect of sex on productivity. In S. Cole and Fiorentine (1991) I offer a "normative alternatives" theory to explain the productivity gap.
6. For a recent discussion and review of the literature on the concept of accumulative advantage see Zuckerman (1989b).

7. The importance of both self- and social selection in determining where scientists work and study is analyzed in detail in Zuckerman (1977a). See also J. R. Cole and S. Cole (1973) and Allison and Long (1987).

8. Such a study would include data on multiple offers to the same individual, with the offer being the unit of analysis. It should also compare those being offered jobs with a random sample of those not being made offers.

9. Allison and Long (1987) do show that productivity had a statistically significant, though weak, influence on the prestige of the destination department.

10. Allison (private communication) reports that the variability in publications and citations among job changers is almost as great as in the full sample. But it would be informative to see a table for professors who are being hired at the tenure level showing productivity by prestige of destination department. It is possible that the variation among job changers could fall around a significantly higher mean.

11. In order to disguise the identity of the scientists involved in some of the cases I have changed descriptions of the individuals involved and the contexts.

12. For a discussion of the difficulty in interpreting even this case, see S. Cole, Rubin, and J. R. Cole (1978), pp. 102–106.

13. For a detailed analysis of how particularistic criteria of evaluation have been used in two areas of biology see Hull (1988).

14. There has been a substantial amount of empirical research by social psychologists on why people in general like or dislike each other (Aronson, 1988), but no research on why scientists like each other. It would not be surprising to find some of the same factors influencing the personal valences of scientists that influence the personal valences of other people. People like those who agree with them or praise them more than they like those who disagree with them or criticize them.

8. Conceptualizing and Studying Particularism in Science

1. Even this classification would in practice not be so clear cut. Some scientists might argue that it is not particularistic to use criteria such as race or gender as part of affirmative action programs aimed at rectifying past discrimination. According to the logic employed in my prior work and in this book, such behavior would constitute particularism if the irrelevant statuses were used in making decisions on, as opposed to searching for, qualified applicants.

2. Here, too, there would be disagreement; some would argue that "collegiality" is a legitimate functionally relevant criterion for the distribution of certain rewards, such as appointment to an academic department.

3. I use sociology because it is the field that I am the most familiar with. As is clear from the data presented in Chapters 3 and 4, high levels of disagreement exist in the natural sciences as well as in the social sciences.

4. J. R. Cole (1987) tries to deal with this problem by making a distinction between whether the evaluation is made on the basis of cognitive labels

or on the basis of a detailed evaluation of the actual content: "If the reviewer rejects the proposal without review because he 'knows' its content by virtue of a label he has affixed to the applicant, or he accepts the proposal on the basis of 'authority' that 'tells' him that the applicant is 'good,' this represents particularism based upon the applicant's membership, as Parsons put it, in a solidary group, clan, or 'invisible college.' However, if the proposal is rejected on the basis of a detailed review of the actual content, and is accompanied by a set of explicit reasons for the decision, which the reviewer applies uniformly to all applications, then the judgment is universalistic" (pp. 3–4).

5. For an informative critical review of the research attempting to test this hypothesis see Zuckerman (1988), pp. 517–519.

6. Some feminist scholars (Harding, 1986; Keller, 1985) have argued that science is inherently biased against women in the way in which problems have been defined and solved. For a critical review see Levin (1988).

7. Several senior scientists who have been involved in the evaluation system over a long period of time commented to me that the proportion of critical letters seems to have declined in recent years and that in general there is less trust that evaluators will say what they really think about the work they are being asked to evaluate. If this is indeed true it may be a result of both growth in the size of interpersonal networks and the linking of networks that in the past may have been relatively autonomous. Today, for example, because of the high mobility among senior scientists, it is possible that they are more concerned with maintaining the good will of colleagues all over the world than they were in the past, when mobility was less common.

8. Would using the prestige or selectivity of the applicant's undergraduate institution be particularistic? A positive answer to this question would require us to make the assumption that GRE scores enable us to partial out all the ability, achievement, and motivational differences between graduates of selective and nonselective schools. This is equivalent to Allison, Long, and McGinnis' assumption that predoctoral productivity enables us to partial out all talent differences between graduates of prestigious and less prestigious Ph.D. departments (see Chapter 7).

9. The tendency to use particularistic criteria in hiring members of one's own department is increased because the new member can have an influence on the everyday life of the evaluator. A newly elected member of the SRA will have no influence on the elector.

10. It would, in fact, be difficult to design any study which would allow us to make such a determination. We could not ask the evaluators to tell us the bases of their judgments because norms prescribing universalistic evaluation might influence their responses. Also many evaluators may not be consciously aware of the bases of their evaluations. We could find out the extent to which network ties were correlated with evaluations but it would be difficult or impossible to determine the bases of the ties.

11. Some of these characteristics, such as gender, have a significant influence in determining who goes into science.

12. For an important analysis of the role of decentralization in the development of science see Ben-David (1960, 1991).

9. Social Influences on the Rate of Scientific Advance

1. The data in this chapter were previously discussed in S. Cole and Meyer (1985). An earlier version of this chapter was presented to the Polish Academy of Sciences in May 1987. The research reported here is based upon Garry Meyer's (1979) doctoral dissertation, and he conducted all the empirical analysis reported in this chapter. I thank him for his help. I thank Eugene Garfield and Henry Small of the Institute for Scientific Information for providing the raw data underlying the analysis reported in this chapter. The discussion of the work of Merton and Ben-David in this chapter is to some extent a rational reconstruction, because some of the ideas attributed to them are implicit rather than explicit.
2. There is a very high correlation between the amount of electricity consumed and GNP.
3. Price does not report the exact correlation.
4. Many researchers who have analyzed national differences in scientific performance have pointed out that the *Science Citation Index* overrepresents English-language journals and underrepresents Russian-language journals. Nonetheless, when analyses based upon a larger number of journals are compared with those based upon a smaller number the results are generally equivalent (Irvine and Martin, 1989). There may also be a problem in the overestimation of the GNP of the Soviet Union, which could explain the smaller number of scientific publications than expected.
5. Because the *DNB* lists only "successful" scientists (those who succeeded in making important discoveries), it is possible that the total number of talented people interested in science throughout the century did not change, but that the number of those successful in making important discoveries did. This again brings up the important question of the extent to which the number of important discoveries is a linear function of the number of scientists. If there is such a linear relationship, then there is no problem in the interpretation Merton has given the data.
6. For a recent discussion of the Merton thesis see Cohen (1990).
7. Merton first introduced the concept of "opportunity structure" in his article "Social Structure and Anomie" (1957b).
8. Whether Merton's empirical generalization is correct is a question different from the one of whether his *explanation* of this generalization is correct. Merton suggests that the important discoverers attract others to science, thus creating the correlation, but it is possible that both the important and unimportant discoverers could be drawn to science by other variables.
9. See, for examples, pp. 6, 38, 39, 46, 53–55, 59, 102, and 103.
10. It is interesting to consider comparing the rate of growth in frontier

knowledge and core knowledge. It is possible that the rate of growth of the core might be limited in a similar way to the rate of growth in the number of Nobel laureates. If the size of the core is restricted, then as the frontier grows, the size of the core will grow at a slower rate, but the "quality" of the contributions entering the core might increase.

11. In fact a few areas of mathematics may depend upon increasing use of resources, such as large computers. Given the fact that the cost to power ratio has been going down in computers, however, even these fields should not face serious resource barriers.

12. Rescher (1978) argues that although it might be difficult to measure conceptual progress, we can measure the power of science to influence the world. He may be correct, but given that many scientific discoveries never have any practical applications and that there is frequently a substantial time lag between a discovery and its application, this criterion would be difficult or impossible to use to determine the importance of scientific discoveries made in ten- or even twenty-five-year periods.

13. Rescher (1978, p. 109) assumes without data that the Ortega hypothesis is correct. Because he believes it requires more and more resources to make important discoveries equivalent to those made in the past, he faces the problem of whether the work of the large number of scientists who do not make the important discoveries is of any value for scientific progress. He assumes that their work is "useful and necessary for genuine advances, the indispensable grist, so to speak, for the mill of scientific progress."

14. This conclusion is still challenged by some. The Ortega hypothesis was the topic of an entire issue of *Scientometrics* (vol. 12, nos. 5–6, 1987, pp. 267–429).

15. The field of physics was selected for this study in part because the qualitative evidence suggested that academic physics had indeed gone through a substantial reduction in demand. For a complete discussion see Meyer (1979). Caution should be exercised in generalizing from this case study to other fields and other social settings.

16. To count how many new assistant professors were hired by Ph.D.-granting departments from 1963 to 1976, we entered information on every physicist listed in the American Institute of *Physics Guide to Faculties and Staff* for the years 1962 to 1976 into a computerized data base. We then used these data to identify each new assistant professor hired between 1963 and 1976. For a full description of the procedure see Meyer (1979), p. 20.

17. For a more detailed discussion of the validity of citations as a rough measure of relative quality see J. R. Cole and S. Cole (1973), chap. 2; Narin (1976); Edge (1979); Meyer (1979). For a discussion of using citations to evaluate individuals as opposed to groups of scientists see S. Cole (1989).

18. Meyer (1979, p. 80) found that out of the 41,540 people who were cited at least once in either *Physical Review* or *Physical Review Letters* in 1967, 1971, and 1977, there were only 10 to 20 people (less than half of one-

tenth of a percent) who received less than four citations to their work in the 1967 and 1971 editions of the *Science Citation Index* and four or more citations to their work in the 1977 edition.

19. There is no reason to expect that decline in productivity would be greater in any particular cohort.

20. In 1967 there were 76,823 citations made in these two journals and in 1977 a total of 94,595 citations (Meyer, 1979, p. 169).

21. A detailed discussion of the problems involved in using the citation indices as output measures may be found in Meyer (1979). One limitation on the data presented here is that they represent only citations to single-authored or first-authored work. The correlation between citations to single-authored or first-authored work and citations to all work is generally very high (see Chapter 5). Nonetheless, this work underestimates the proportion of physicists whose work received citation. It is unlikely, however, that this would have had differential effects on the cohorts we studied.

22. Using Rescher's notion of truly significant discoveries, it is probable that only a handful of the scientists in our entire sample would qualify. Because there has been an exponential growth in the number of scientists, Rescher's conclusion becomes true by definition if we limit the definition of "significant" discovery severely enough.

23. The work of some of the cohorts was slightly older than that of others. The work of the 1963 cohort, for example, was older than the work of the 1964 cohort. Given, however, the lenient criterion we are using in determining whether or not a scientist is making a contribution, it is unlikely that these age differences had any significant substantive effect. Nonetheless this work should be replicated using citation data from every year in order to control for age differences in the literature. We were able to do this in making specific comparisons, such as that between the 1965 cohort, the 1969 cohort, and the 1975 cohort. Whenever we made such comparisons we found confirmation of our general conclusion.

24. It has often been hypothesized that the making of an important discovery will bring about other important discoveries. See Merton (1970), pp. 43, 48, and Rescher (1978), pp. 211–212.

10. The Future of the Sociology of Science

1. It is true that once a new contribution is published it can be taken up and modified by others in the community. Local knowledge outcomes are influenced by the existing state of knowledge and the work of others in the community. But a modification would result in a new local knowledge outcome whose substance would be determined in the local context, with the work of others as one influencing variable.

2. For a reply to this type of attack see Ashmore (1989).

3. She does show how general cognitive characteristics of the paper, such as its discussion of the methods and techniques used, changed from draft to draft.

Appendix

1. For a more detailed account see J. R. Cole and S. Cole (1981). For a description of the first phase see S. Cole, Rubin, and J. R. Cole (1978).
2. For the names of the chairs and members of the reviewer-selector panels see J. Cole and S. Cole (1981), pp. 9–10.
3. The variables used to characterize principal investigators in Phase I were measured in the same way, with exceptions noted.
4. For Phase I we counted citations in 1974 to work published between 1965 and 1974.
5. For Phase I we counted citations in 1974 to work published prior to 1965.
6. For Phase I we counted all papers published between 1965 and 1974.
7. For Phase I this was the number of years between 1970 and 1974 in which the principal investigator had received funds from the NSF.
8. The variable was not included in the analysis of the Phase II data.
9. For Phase I those who obtained their degrees in 1970 or later were coded as 0, and those receiving their degrees in 1969 or earlier were coded as 1.
10. This model was designed by Lee Cronbach, Jack Kieffer, and Gary Simon and run by Gary Simon. I thank them for their help.

References

Adatto, Kiku, and Stephen Cole. 1981. "The Functions of Classical Theory in Contemporary Sociological Research: The Case of Max Weber." In Robert Alun Jones and Henrika Kuklick, eds., *Knowledge and Society: Studies in the Sociology of Culture Past and Present*, pp. 137–162. Greenwich, Conn.: JAI Press.

Allison, Paul D., and J. Scott Long. 1987. "Inter-University Mobility of Academic Scientists." *American Sociological Review* 52: 643–652.

——— 1990. "Departmental Effects on Scientific Productivity." *American Sociological Review* 55: 469–478.

Allison, Paul D., and John A. Stewart. 1974. "Productivity Differences among Scientists: Evidence for Accumulative Advantage." *American Sociological Review* 39: 596–606.

Aronson, Elliot. 1988. *The Social Animal*, 5th ed., chap. 7. New York: W. H. Freeman.

Aronson, Naomi. 1984. "The Discovery of Resistance: Historical Accounts and Scientific Careers." Paper presented at American Sociological Association meetings, Austin, Texas.

Ashmore, Malcolm. 1989. *The Reflexive Thesis: Wrighting Sociology of Scientific Knowledge*. Chicago: University of Chicago Press.

Bakanic, Von, Clark McPhail, and Rita J. Simon. 1987. "The Manuscript Review and Decision-Making Process." *American Sociological Review* 52: 631–642.

Barber, Bernard. 1961. "Resistance by Scientists to Scientific Discovery." *Science* 134: 596–602.

Barnes, Barry. 1974. *Scientific Knowledge and Sociological Theory*. London: Routledge and Kegan Paul.

——— 1977. *Interests and the Growth of Knowledge*. London: Routledge and Kegan Paul.

Barnes, S. B., and R. G. A. Dolby. 1970. "The Scientific Ethos: A Deviant Viewpoint." *European Journal of Sociology* 2: 3–25.

Barnes, Barry, and Donald MacKenzie. 1979. "On the Role of Interests in Scientific Change." In Roy Wallis, ed., *On the Margins of Science: The Social Construction of Rejected Science*, pp. 49–66. Sociological Review Monographs. Keele: University of Keele.

Ben-David, Joseph. 1960. "Scientific Productivity and Academic Organization in Nineteenth-Century Medicine." *American Sociological Review* 25: 828–843.

——— 1971. *The Scientist's Role in Society: A Comparative Study*. Englewood Cliffs, N.J.: Prentice-Hall.

——— 1991. *Scientific Growth: Essays on the Social Organization and Ethos of Science*. Berkeley: University of California Press.

Ben-David, Joseph, and Awraham Zloczower. 1962. "Universities and Academic Systems in Modern Society." *European Journal of Sociology* 3: 45–84.

Berger, Peter L., and Thomas Luckmann. 1963. *The Social Construction of Reality*. New York: Doubleday.

Beyer, Janice M. 1978. "Editorial Policies and Practices among Leading Journals in Four Scientific Fields." *Sociological Quarterly* 19: 68–88.

Bickel, P. J., E. A. Hammel, and J. W. O'Connell. 1975. "Sex Bias in Graduate Admissions: Data from Berkeley." *Science* 187: 398–404.

Bloor, David. 1976. *Knowledge and Social Imagery*. London: Routledge and Kegan Paul.

——— 1982. "Durkheim and Mauss Revisited: Classification and the Sociology of Knowledge." *Studies in History and Philosophy of Science* 13: 267–297.

Brannigan, Augustine. 1981. *The Social Basis of Scientific Discovery*. Cambridge: Cambridge University Press.

Breneman, David. 1975. "Outlook and Opportunity for Graduate Education." Technical Report no. 3 of the National Board of Graduate Education, Washington, D.C.

Brush, Stephen. 1974. "Should the History of Science Be Rated X?" *Science* 183: 1164–72.

Caplow, Theodore, and Reece J. McGee. 1958. *The Academic Marketplace*. New York: Basic Books.

Carter, Grace M. 1974. *Peer Review, Citations, and Biomedical Research Policy: NIH Grants to Medical School Faculty*. Santa Monica: Rand (R-1583-HEW).

Cartter, Alan. 1976. *Ph.D.'s and the Academic Labor Market*. New York: McGraw-Hill.

Chubin, Daryl E., and Edward Hackett. 1990. *Peerless Science: Peer Review and U.S. Science Policy*. Albany: State University of New York Press.

Chubin, Daryl E., and Soumyo Moitra. 1975. "Content Analysis of References: An Alternative to Citation Counting." *Social Studies of Science* 5: 423–441.

Cicchetti, Domenic V. 1991. "The Reliability of Peer Review for Manuscript and Grant Submissions: A Cross-Disciplinary Investigation." *Behavioral and Brain Sciences* 14: 119–186.

Cohen, I. Bernard, ed. 1990. *Puritanism and the Rise of Modern Science: The Merton Thesis*. New Brunswick: Rutgers University Press.

Cole, Jonathan R. 1979. *Fair Science*. New York: Free Press.

——— 1987. "The Paradox of Individual Particularism and Institutional Universalism." In J. Elster, ed., *Justice and the Lottery*. Cambridge: Cambridge University Press.

Cole, Jonathan R., and Stephen Cole. 1972. "The Ortega Hypothesis." *Science* 178: 368–375.

——— 1973. *Social Stratification in Science*. Chicago: University of Chicago Press.

——— 1979. "Which Researchers Get the Grants." *Nature* 279: 575–576.

——— with the Committee on Science and Public Policy, National Academy of Sciences. 1981. *Peer Review in the National Science Foundation: Phase Two of a Study*. Washington, D.C.: National Academy Press.

——— 1985. "Experts' Consensus and Decision Making at the National Science Foundation." In Kenneth S. Warren, ed., *Selectivity in Information Systems: Survival of the Fittest*, pp. 27–63. New York: Praeger Science Publishers.

Cole, Stephen. 1970. "Professional Standing and the Reception of Scientific Discoveries." *American Journal of Sociology* 76: 286–306.

——— 1975. "The Growth of Scientific Knowledge: Theories of Deviance as a Case Study." In Lewis A. Coser, ed., *The Idea of Social Structure: Papers in Honor of Robert K. Merton*, pp. 175–220. New York: Harcourt Brace Jovanovich.

——— 1978. "Scientific Reward Systems: A Comparative Analysis." In R. A. Jones, ed., *Research in Sociology of Knowledge, Sciences and Art*. pp. 167–190. Greenwich, Conn.: JAI Press.

——— 1979. "Age and Scientific Performance." *American Journal of Sociology* 84: 958–977.

——— 1983. "The Hierarchy of the Sciences?" *American Journal of Sociology* 89: 111–139.

——— 1986. "Sex Discrimination and Admission to Medical School: 1929–1984." *American Journal of Sociology* 92: 549–567.

——— 1989. "The Use of Citations in the Evaluation of Individual Scientists." *Trends in Biochemical Research* 14: 9–12.

Cole, Stephen, and Jonathan R. Cole. 1967. "Scientific Output and Recognition." *American Sociological Review* 32: 377–390.

Cole, Stephen, Jonathan R. Cole, and Lorraine Dietrich. 1978. "Measuring the Cognitive State of Scientific Disciplines." In Y. Elkana, J. Lederberg, R. K. Merton, A. Thackray, and H. Zuckerman, eds., *Toward a Metric of Science: The Advent of Science Indicators*, pp. 209–251. New York: Wiley.

Cole, Stephen, Jonathan R. Cole, and Gary Simon. 1981. "Chance and Consensus in Peer Review." *Science* 214: 881–886.

Cole, Stephen, and Robert Fiorentine. 1991. "Discrimination against Women in Science: The Confusion of Outcome with Process." In H. Zuckerman, J. R. Cole, and J. Bruer, eds., *The Outer Circle: Women in the Scientific Community*, pp. 205–226. New York: W. W. Norton.

Cole, Stephen, and Garry S. Meyer. 1985. "Little Science, Big Science Revisited." *Scientometrics* 7: 443–458.

Cole, Stephen, Leonard Rubin, and Jonathan R. Cole. 1978. *Peer Review in the National Science Foundation: Phase I*. Washington, D.C.: National Academy of Sciences.

Cole, Stephen, Gary Simon, and Jonathan R. Cole. 1988. "Do Journal Rejection Rates Index Consensus?" *American Sociological Review* 53: 152–156.

Cole, Stephen, and Harriet Zuckerman. 1976. "The Use of ACE Ratings in Research on Science and Higher Education." Prepared for planning conference on assessment of the quality of graduate education programs in the United States, September 27–29, Woods Hole, Massachusetts.

Collins, Harry M. 1974. "The TEA-set: Tacit Knowledge and Scientific Networks." *Science Studies* 4: 165–186.

——— 1975. "The Seven Sexes: A Study in the Sociology of a Phenomenon, or the Replication of Experiments in Physics." *Sociology* 9: 205–224.

——— 1981. "Stages in the Empirical Program of Relativism." *Social Studies of Science* 11: 3–10.

——— 1982. "Knowledge, Norms, and Rules in the Sociology of Science." *Social Studies of Science* 12: 299–309.

——— 1985. *Changing Order: Replication and Induction in Scientific Practice.* London: Sage.

Collins, Harry M., and Trevor Pinch. 1982. *Frames of Meaning: The Social Construction of Extraordinary Science.* London: Routledge and Kegan Paul.

Collins, Randall. 1986. "Is Sociology in the Doldrums?" *American Journal of Sociology* 91: 1336–55.

Conant, James B. 1950. Foreword to *Harvard Case Studies in Experimental Science.* Cambridge: Harvard University Press.

Crane, Diana. 1965. "Scientists at Major and Minor Universities: A Study in Productivity and Recognition." *American Sociological Review* 30: 699–714.

——— 1980. "An Exploratory Study of Kuhnian Paradigms in Theoretical High Energy Physics." *Social Studies of Science* 10: 23–54.

Crick, Francis. 1988. *What Mad Pursuit: A Personal View of Scientific Discovery.* New York: Basic Books.

Durkheim, Emile. [1890] 1951. *Suicide.* Glencoe, Ill.: Free Press.

Edge, David. 1979. "Quantitative Measures of Communication in Science: A Critical Review." *History of Science* 8: 102–134.

Elkana, Yehuda. 1970. "The Conservation of Energy: A Case of Simultaneous Discovery." *Archives Sciences* 23: 31–60.

——— 1978. "Two-Tier-Thinking: Philosophical Realism and Historical Relativism." *Social Studies of Science* 8: 309–326.

——— 1981. "A Programmatic Attempt at an Anthropology of Knowledge." In Everett Mendelsohn and Yehuda Elkana, eds., *Science and Cultures,* pp. 1–76. Dordrecht: D. Reidel.

Feyerabend, Paul. 1975. *Against Reason.* London: NLB.

Fiorentine, Robert. 1987. "Men, Women, and the Premed Persistence Gap: A Normative Alternatives Approach." *American Journal of Sociology* 92: 1118–39.

——— 1988. "Increasing Similarity in the Values and Life Plans of Male and Female College Students? Evidence and Implications." *Sex Roles* 18: 143–158.

Fiorentine, Robert and Stephen Cole. 1992. "Why Fewer Women Become Physicians: Explaining the Premed Persistence Gap." *Sociological Forum* (in press).

Freudenthal, Gad. 1984. "The Role of Shared Knowledge in Science: The Failure of the Constructivist Programme in the Sociology of Science." *Social Studies of Science* 14: 285–295.

Galison, Peter. 1987. *How Experiments End.* Chicago: University of Chicago Press.

Garber, Elizabeth. 1989. "Introduction." In Elizabeth Garber, ed., *Beyond History of Science: Essays in Honor of Robert E. Sihofill,* pp. 7–20. Bethlehem, Pa.: Lehigh University Press.

Garber, Elizabeth, and Fred Weinstein. 1987. "History of Science as Social History." In Jerome Rabow, Gerald M. Platt, and Marion S. Goldman, eds., *Advances in Psychoanalytic Sociology,* pp. 279–298. Malabar, Fl.: Krieger.

Garfinkel, Harold, Michael Lynch, and Eric Livingston. 1981. "The Work of a Discovering Science Construed with Materials from the Optically Discovered Pulsar." *Philosophy of the Social Sciences* 11:131–158.

Garvey, William D., Nan Lin, and Carnot E. Nelson. 1970. "Some Comparisons of Communication Activities in the Physical and Social Sciences." In Carnot E. Nelson and Donald K. Pollock, eds., *Communication among Scientists and Engineers,* pp. 61–84. Lexington, Mass.: Heath.

Gaston, Jerry. 1973. *Originality and Competition in Science.* Chicago: University of Chicago Press.

―――― 1978. *The Reward System in British and American Science.* New York: Wiley.

Giere, Ronald N. 1988. *Explaining Science.* Chicago: University of Chicago Press.

Gieryn, Thomas. 1982. "Relativist/Constructivist Programmes in the Sociology of Science: Redundance and Retreat." *Social Studies of Science* 12: 279–297.

Gilbert, G. Nigel, and Michael Mulkay. 1984. *Opening Pandora's Box: A Sociological Analysis of Scientists' Discourse.* Cambridge: Cambridge University Press.

Gingras, Yves, and Silvan S. Schweber. 1986. "Constraints on Construction" (review of Pickering, *Constructing Quarks*). *Social Studies of Science* 16: 372–383.

Gleick, James. 1987. *Chaos: Making a New Science.* New York: Viking.

Granovetter, Mark. 1973. "The Strength of Weak Ties." *American Journal of Sociology* 78: 1360–80.

Grodzins, Lee. 1976. *Physics Faculties, 1959–1975.* Mimeographed.

Haberfeld, Yitchak, and Yehouda Shenhav. 1988. "Are Women and Blacks Closing the Gap? Salary Discrimination in American Science during the 1970s and the 1980s." *Industrial and Labor Relations Review* 44: 68–82.

Hagstrom, Warren O. 1971. "Inputs, Outputs, and the Prestige of University Science Departments." *Sociology of Education* 44: 375–397.

―――― 1974. "Competition in Science." *American Sociological Review* 39: 1–18.

Harding, Sandra. 1986. *The Science Question in Feminism.* Ithaca: Cornell University Press.

Hargens, Lowell L. 1975. *Patterns of Scientific Research: A Comparative Analysis of Research in Three Scientific Fields.* Washington, D.C.: American Sociological Association.

―――― 1988a. "Cognitive Consensus and Journal Rejection Rates." *American Sociological Review* 53: 139–151.

——— 1988b. "Further Evidence on Field Differences in Consensus from the NSF Peer Review Studies." *American Sociological Review* 53: 157–160.

Hargens, Lowell L., and Warren O. Hagstrom. 1967. "Sponsored and Contest Mobility of American Academic Scientists." *Sociology of Education* 40: 24–38.

——— 1982. "Scientific Consensus and Academic Status Attainment Patterns." *Sociology of Education* 55: 183–196.

Harvey, Bill. 1980. "The Effects of Social Context on the Process of Scientific Investigation: Experimental Tests of Quantum Mechanics." In Karin D. Knorr-Cetina, Roger Krohn, and Richard Whitley, eds., *The Social Process of Scientific Investigation*, pp. 139–163. Dordrecht: D. Reidel.

Hessen, Boris. [1931] 1971. "The Social and Economic Roots of Newton's *Principia*." In *Science at the Crossroads: Papers Presented to the International Congress of the History of Science and Technology*, pp. 149–212. London: Frank Cass & Co.

Holmes, Frederic L. 1987. "Scientific Writing and Scientific Discovery." *Isis* 78: 220–235.

Horowitz, Lawrence C. 1988. *Taking Charge of Your Medical Fate*. New York: Random House.

Hull, David. 1988. *Science as Process*. Chicago: University of Chicago Press.

Irvine, J. E., and B. R. Martin. 1989. "International Comparisons of Scientific Performance Revisited." *Scientometrics* 15: 369–392.

Jencks, Christopher, et al. 1972. *Inequality: A Reassessment of the Effects of Family and Schooling in America*. New York: Basic Books.

Keller, Evelyn Fox. 1985. *Reflections on Gender and Science*. New Haven: Yale University Press.

King, M. D. 1971. "Reason, Tradition, and the Progressiveness of Science." *History and Theory* 10: 3–32.

Klitgaard, Robert E. 1979. "The Decline of the Best? An Analysis of the Relationships between Declining Enrollments, Ph.D. Production, and Research." John F. Kennedy School of Government, Harvard University, discussion paper 65D.

Knorr-Cetina, Karin D. 1975. "The Nature of Scientific Consensus and the Case of the Social Sciences." In Karin D. Knorr, Herman Strasser, and Hans Georg Zilian, eds., *Determinants and Controls of Scientific Development*, pp. 227–256. Dordrecht: D. Reidel.

——— 1981. *The Manufacture of Knowledge: An Essay on the Constructivist and Contextual Nature of Science*. New York: Pergamon Press.

——— 1982. "The Constructivist Programme in the Sociology of Science: Retreats or Advances?" *Social Studies of Science* 12: 320–324.

——— 1983."The Ethnographic Study of Scientific Work: Towards a Constructivist Interpretation of Science." In Karin D. Knorr-Cetina and Michael Mulkay, eds., *Science Observed*. London: Sage.

Kuhn, Thomas S. [1962] 1970. *The Structure of Scientific Revolutions*, 2nd ed. Chicago: University of Chicago Press.

——— [1961] 1977. "The Function of Measurement in Modern Physical Science." In Thomas S. Kuhn, *The Essential Tension*, pp. 178–224. Chicago: University of Chicago Press.

Lakatos, Imre. 1970. "Falsification and the Methodology of Research Programmes." In Imre Lakatos and Alan Musgrave, eds., *Criticism and the Growth of Knowledge*, pp. 91–196. Cambridge: Cambridge University Press.

Latour, Bruno. 1980. "Is It Possible to Reconstruct the Research Process? Sociology of a Brain Peptide." In Karin Knorr-Cetina, Roger Krohn, and Richard Whitley, eds., *The Social Process of Scientific Investigation*, pp. 53–73. Dordrecht: D. Reidel.

—— 1987. *Science in Action*. Cambridge: Harvard University Press.

—— 1988. *The Pasteurization of France*. Translated by Alan Sheridan and John Law. Cambridge: Harvard University Press.

Latour, Bruno, and Steve Woolgar. [1979] 1986. *Laboratory Life: The Construction of Scientific Facts*. Princeton: Princeton University Press.

Laudan, Larry. 1977. *Progress and Its Problems: Toward a Theory of Scientific Growth*. Berkeley: University of California Press.

—— 1984. *Science and Values: The Aims of Science and Their Role in Scientific Debate*. Berkeley: University of California Press.

—— 1990. *Relativism and Science*. Chicago: University of Chicago Press.

Levin, Margarita. 1988. "Caring New World: Feminism and Science." *American Scholar* 57: 100–106.

Lieberson, Stanley. 1985. *Making It Count: The Improvement of Social Research and Theory*. Berkeley: University of California Press.

Lodahl, Janice B., and Gerald Gordon. 1972. "The Structure of Scientific Fields and the Functioning of University Graduate Departments." *American Sociological Review* 37: 57–72.

Loftus, Elizabeth. 1979. *Eyewitness Testimony*. Cambridge: Harvard University Press.

Long, J. Scott. 1978. "Productivity and Position in the Scientific Career." *American Sociological Review* 43: 899–908.

Long, J. Scott, Paul D. Allison, and Robert McGinnis. 1979. "Entrance into the Academic Career." *American Sociological Review* 44: 816–830.

Long, J. Scott, and Robert McGinnis. 1981. "Organizational Context and Scientific Productivity." *American Sociological Review* 46: 422–442.

Lynch, Michael. 1985. *Art and Artifact in Laboratory Science*. London: Routledge & Kegan Paul.

MacKenzie, Donald. 1981. *Statistics in Britain, 1865–1930: The Social Construction of Scientific Knowledge*. Edinburgh: Edinburgh University Press.

Mannheim, Karl. [1929] 1954. *Ideology and Utopia*. New York: Harcourt, Brace.

Masterman, Margaret. 1970. "The Nature of a Paradigm." In Imre Lakatos and Alan Musgrave, eds., *Criticism and the Growth of Knowledge*, pp. 59–89. Cambridge: Cambridge University Press.

Medawar, Peter. 1963. "Is the Scientific Paper a Fraud?" *The Listener* (12 Sept.): 377–378.

Merton, Robert K. 1937. "Science, Population, and Society." *Scientific Monthly* 44: 165–171.

—— [1942] 1957a. "Science and Democratic Social Structure." In Robert K. Merton, *Social Theory and Social Structure*, rev. ed., pp. 550–561. Glencoe, Ill.: Free Press. Also reprinted in Merton, 1973.

—— [1938] 1957b. "Social Structure and Anomie." In Robert K. Merton, *Social Theory and Social Structure*, rev. ed., pp. 131–160. Glencoe, Ill.: Free Press.

—— 1968. "The Matthew Effect in Science." *Science* 159: 56–63. Reprinted in Merton, 1973.

—— [1938] 1970. *Science, Technology, and Society in Seventeenth-Century England*. New York: Howard Fertig.

—— 1973. *The Sociology of Science: Theoretical and Empirical Investigations*. Edited by Norman W. Storer. Chicago: University of Chicago Press.

—— 1976. *Sociological Ambivalence and Other Essays*. New York: Free Press.

—— 1988. "The Matthew Effect in Science, II: Cumulative Advantage and the Symbolism of Intellectual Property." *Isis* 79: 606–623.

—— 1990. "STS: Foreshadowings of an Evolving Research Program in the Sociology of Science." In I. Bernard Cohen, ed., *Puritanism and the Rise of Modern Science: The Merton Thesis*, pp. 334–371. New Brunswick: Rutgers University Press.

Meyer, Garry Stuart. 1979. "Academic Labor and the Development of Science." Ph.D. dissertation, State University of New York at Stony Brook.

Mitroff, Ian I. 1974. *The Subjective Side of Science*. Amsterdam: Elsevier.

Moravcsik, Michael J., and Poovanalingam Murugesan. 1975. "Some Results on the Function and Quality of Citations." *Social Studies of Science* 5: 86–92.

Mulkay, Michael. 1977. "The Sociology of Science in Britain." In Robert K. Merton and Jerry Gaston, eds., *The Sociology of Science in Europe*. Carbondale: Southern Illinois University Press.

—— 1978. "Consensus in Science." *Social Science Information* 17, 1, 107–122.

—— 1979. *Science and the Sociology of Knowledge*. London: George Allen and Unwin.

Mulkay, Michael, and A. T. Williams. 1971. "A Sociological Study of a Physics Department." *British Journal of Sociology* 22: 68–82.

Mullins, Nicholas, Lowell L. Hargens, Pamela Hecht, and Edward Kick. 1977. "The Group Structure of Cocitation Clusters: A Comparative Study." *American Sociological Review* 42: 552–563.

Nagel, Ernest. 1961. *The Structure of Science*. New York: Harcourt, Brace, and World.

Narin, Francis. 1976. *Evaluative Bibliometrics*. Cherry Hill, N.J.: Computer Horizons.

National Science Foundation. 1976. *Peer Review*. Washington, D.C.

—— 1977. *Science Indicators 1976*. Washington, D.C.: U.S. Government Printing Office.

Oromaner, Mark. 1972. "The Structure of Influence in Contemporary Academic Sociology." *American Sociologist* 7:11–13.

Pfeffer, Jeffrey, Gerald R. Salancik and Huseyin Leblebici. 1976. "The Effect of Uncertainty on the Use of Social Influence in Organizational Decision Making." *Administrative Science Quarterly* 21: 227–245.

Pickering, Andrew. 1980. "The Role of Interests in High-Energy Physics: The Choice between Charm and Colour." In Karin D. Knorr-Cetina, Roger Krohn, and Richard Whitley, eds., *The Social Process of Scientific Investigation*, pp. 107–138. Dordrecht: D. Reidel.

—— 1984. *Constructing Quarks: A Sociological History of Particle Physics*. Edinburgh: Edinburgh Unversity Press.

—— 1989. "Living in the Material World." In David Gooding, Trevor Pinch, and Simon Schaffer, eds., *The Uses of Experiment*, pp. 275–297. Cambridge: Cambridge University Press.

Pinch, Trevor J. 1980. "Theoreticians and the Production of Experimental Anomaly: The Case of Solar Neutrinos." In Karin D. Knorr-Cetina, Roger Krohn, and Richard Whitley, eds., *The Social Process of Scientific Investigation*. pp. 77–106. Dordrecht: D. Reidel.

—— 1986. *Confronting Nature: The Sociology of Solar-Neutrino Detection*. Dordrecht: D. Reidel.

Polanyi, Michael. 1958. *Personal Knowledge*. London: Routledge & Kegan Paul.

—— 1963. "The Potential Theory of Adsorption." *Science* 141: 1010–13.

Potter, Jonathan. 1984. "Testability, Flexibility: Kuhnian Values in Scientists' Discourse Concerning Theory Choices." *Philosophy of the Social Sciences* 14: 303–330.

Price, Derek J. de Solla. 1963. *Little Science, Big Science*. New York: Columbia University Press.

—— [1969] 1986. "Measuring the Size of Science." In Derek J. de Solla Price, *Little Science, Big Science . . . and Beyond*, pp. 135–154. New York: Columbia University Press.

Price, Derek J. de Solla, and Suha Gursey. 1986. "Some Statistical Results for the Numbers of Authors in the States of the United States and the Nations of the World." In Derek J. de Solla Price, *Little Science, Big Science . . . and Beyond*, pp. 180–205. New York: Columbia University Press.

Radner, Roy, and Leonard S. Miller. 1975. *Demand and Supply in United States Higher Education*. Carnegie Commission on Higher Education. New York: McGraw Hill.

Rescher, Nicholas. 1978. *Scientific Progress: A Philosophical Essay on the Economics of Research in Natural Science*. Pittsburgh: University of Pittsburgh Press.

Reskin, Barbara F. 1977. "Scientific Productivity and the Reward Structure of Science." *American Sociological Review* 42: 491–504.

Roose, K. D., and C. J. Anderson. 1971. *A Rating of Graduate Programs*. Washington, D.C.: American Council on Education.

Rubin, Leonard Charles. 1975. "The Dynamics of Tenure in Two Academic Disciplines." Ph.D. dissertation, State University of New York at Stony Brook.

Rudwick, Martin J. S. 1985. *The Great Devonian Controversy: The Shaping of Scientific Knowledge among Gentlemanly Specialists*. Chicago: University of Chicago Press.

Sarton, George. [1927] 1962. *Introduction to the History of Science*. Baltimore: Carnegie Institution.

Shapin, Steven. 1982. "History of Science and Its Sociological Reconstructions." *History of Science* 20: 157–211.

Shapin, Steven, and Simon Schaffer. 1985. *Leviathan and the Air-Pump: Hobbes, Boyle, and the Experimental Life*. Princeton: Princeton University Press.

Shenhav, Yehouda A. and Yitchak Haberfeld. 1988. "Scientists in Organizations: Discrimination Processes in an Internal Labor Market." *Sociological Quarterly* 29: 1–12.

Simonton, Dean Keith. 1988. "Age and Outstanding Achievement: What Do We Know after a Century of Research?" *Psychological Bulletin* 104: 251–267.

Small, Henry G. 1974. *Characteristics of Frequently Cited Papers in Chemistry.* Final Report on Contract no. NSF-C795. Philadelphia: Institute for Scientific Information.

——— 1985. "The Lives of a Scientific Paper." In Kenneth S. Warren, ed., *Selectivity in Information Systems: Survival of the Fittest,* pp. 83–97. New York: Praeger Science Publishers.

Small, Henry G., and Belver Griffith. 1974. "The Structure of Scientific Literature I: Identifying and Graphing Specialties." *Science Studies* 4: 17–40.

Sorokin, Pitirim A., and Robert K. Merton. 1935. "The Course of Arabian Intellectual Development, 700–1300 A.D.: A Study in Method." *Isis* 22: 516–524.

Stehr, Nico. 1978. "The Ethos of Science Revisited." *Sociological Inquiry* 48: 172–198.

Stern, Nancy. 1978. "Age and Achievement in Mathematics." *Social Studies of Science* 8: 127–140.

Stewart, John A. 1990. *Drifting Continents and Colliding Paradigms: Perspectives on the Geoscience Revolution.* Bloomington: Indiana University Press.

Stinchcombe, Arthur L., and Richard Ofshe. 1969. "On Journal Editing as a Probabilistic Process." *American Sociologist* 4: 116–117.

Storer, Norman. 1967. "The Hard Sciences and the Soft: Some Sociological Observations." *Bulletin of the Medical Library Association* 55: 75–84.

Sullivan, Daniel, D. Hywel White, and Edward J. Barboni. 1977. "The State of a Science: Indicators in the Specialty of Weak Interactions." *Social Studies of Science* 7: 167–200.

Summey, Pamela. 1987. "Ideology and Medical Care: The Case of Cesarean Delivery." Ph.D. dissertation, State University of New York at Stony Brook.

Taubes, Gary. 1986. *Nobel Dreams.* New York: Random House.

Watson, James D. 1968. *The Double Helix.* New York: Atheneum.

Whitley, Richard D. 1972. "Black Boxism and the Sociology of Science: A Discussion of the Major Developments in the Field." In P. Halmos, ed., *The Sociology of Science,* Sociological Review Monograph no. 18, pp. 61–92. Keele: University of Keele.

——— 1984. *The Intellectual and Social Organization of the Sciences.* Oxford: Clarendon Press.

Winstanley, Monica. 1976. "Assimilation into the Literature of a Critical Advance in Molecular Biology." *Social Studies of Science* 6: 545–549.

Woolgar, Steve. 1988. *Science: The Very Idea!* London: Tavistock.

Yearley, Steven. 1981. "Textual Persuasion: The Role of Social Accounting in the Construction of Scientific Arguments." *Philosophy of the Social Sciences* 11: 409–435.

Ziman, John M. 1968. *Public Knowledge: An Essay Concerning the Social Dimension of Science.* Cambridge: Cambridge University Press.

Zuckerman, Harriet. 1977a. *The Scientific Elite.* New York: Free Press.

—— 1977b. "Deviant Behavior and Social Control in Science." In Edward Sagarin, ed., *Deviance and Social Change,* pp. 87–138. Beverly Hills: Sage Publications.

—— 1978. "Theory Choice and Problem Choice in Science." *Sociological Inquiry* 48: 65–95.

—— 1988. "The Sociology of Science." In Neal Smelser, ed., *Handbook of Sociology,* pp. 511–574. Newbury Park, Calif.: Sage.

—— 1989a. "The Other Merton Thesis." *Science in Context* 3: 239–267.

—— 1989b. "Accumulation of Advantage and Disadvantage: The Theory and Its Intellectual Biography." In Carol Mongardini and Simonetta Tabboni, eds., *L'Opera de R. K. Merton e La Sociologia Contemporanea,* pp. 153–176. Genoa: ECIG (Edizioni Culturali Internazionale).

Zuckerman, Harriet, and Robert K. Merton. 1973a. "Age, Aging, and Age Structure in Science." In Robert K. Merton, *The Sociology of Science,* edited by Norman Storer, pp. 497–559. Chicago: University of Chicago Press.

—— [1971] 1973b. "Institutionalization and Patterns of Evaluation in Science." In Robert K. Merton, *The Sociology of Science,* edited by Norman Storer, pp. 460–496. Chicago: University of Chicago Press.

Name Index

Subject Index

accumulative advantage, 137, 140, 165–172, 184, 263n6
  and authority, 195–196
affirmative action, 162, 264n1
age
  influence on getting grants, 145
  and rewards, 163
  and scientific performance, 118–122
AIDS, 26–27, 57, 252n15
ambiguity, tolerance of, 18–19
anomalies, 8, 199
anthropology of science, 36–37
Arabian science, 212
authority, 14, 20–21, 28–30, 50, 55, 69–70, 103–104, 194–198, 229

black-boxing content, 4, 38, 61, 68, 70–72, 79–81, 229, 253n17
British science, seventeenth century, 2–3, 75–76, 208–210

Cesarean section, 62, 255n1
charm-color debate, 66–67, 81
chemiosmotic theory, 17–18, 22
citation
  as measure of impact, 16, 52, 120, 141–142, 221, 223–224, 243–244, 263n2, 268nn21,23
  as persuasion, 14, 45
  concentration of, as measure of consensus, 125–127
class interests, 35, 74–76
classical mechanics, 7, 9

co-citation analysis, 52
codification, 106–108, 117–120, 122, 124, 259n19, 261n24
cognitive norms, 39, 254n8
cold fusion, 183
confirmation, as influence on consensus formation, 43, 46
congruence approach, 76–77
consensus
  as influenced by subject matter, 104
  as necessary for progress, 199–200
  at core and frontier, 15–20, 29, 83–101, 230
  concentration of citations as measure of, 125–127, 260
  constructivists' problems in accounting for, 33–60
  constructivists' view of, 10–14, 82–83, 103, 136
  differences between natural and social sciences, 20, 102–136, 258n5
  diffusionist approach to, 47–53
  formation of, 20, 41, 67, 140
  indirect indicators of, 111–122, 259n19, 260
  influence of authority on, 28–30, 194–198
  influence of cognitive content on, 21–24, 39, 47–48, 52–53, 55, 59, 104, 138–139, 183, 230–231
  influence of scientist characteristics on, 20–21, 28, 42, 52, 103, 119, 137–156